Digital Signatures

Jonathan Katz

Digital Signatures

 Springer

Jonathan Katz
Department of Computer Science
University of Maryland
A.V. Williams Bldg.
College Park, MD 20742
USA
jkatz@cs.umd.edu

ISBN 978-0-387-27711-0   e-ISBN 978-0-387-27712-7
DOI 10.1007/978-0-387-27712-7
Springer New York Dordrecht Heidelberg London

Library of Congress Control Number: 2010927931

Printed on acid-free paper

Springer is part of Springer Science+Business Media (www.springer.com)

To Jill, Abigail, and Rena

Preface

As a beginning graduate student, I recall being frustrated by a general lack of accessible sources from which I could learn about (theoretical) cryptography. I remember wondering: *why aren't there more books presenting the basics of cryptography at an introductory level?* Jumping ahead almost a decade later, as a faculty member my graduate students now ask me: *what is the best resource for learning about (various topics in) cryptography?* This monograph is intended to serve as an answer to these questions — at least with regard to digital signature schemes.[1]

Given the above motivation, this book has been written with a beginning graduate student in mind: a student who is potentially interested in doing research in the field of cryptography, and who has taken an introductory course on the subject, but is not sure where to turn next. Though intended primarily for that audience, I hope that advanced graduate students and researchers will find the book useful as well. In addition to covering various constructions of digital signature schemes in a unified framework, this text also serves as a compendium of various "folklore" results that are, perhaps, not as well known as they should be. This book could also serve as a textbook for a graduate seminar on advanced cryptography; in such a class, I expect the entire book could be covered at a leisurely pace in one semester with perhaps some time left over for excursions into related topics. I hope it will also prove helpful to graduate students and researchers in other fields, such as computer security or mathematics, who want to obtain a more thorough appreciation of digital signatures and known results in this area.

The only real prerequisite for this book is a previous course (at the undergraduate or graduate level) covering the basic foundations of modern cryptography. Specifically, I assume the reader has taken a course whose coverage and treatment of cryptography is similar to that of the textbook *Introduction to Modern Cryptography* [72] that I have co-authored with Yehuda Lindell. Comfortability with formal definitions and proofs is expected, and it is assumed the reader is already familiar with, e.g., the RSA and discrete logarithm problems, and the notion of one-way

---

[1] Fortunately, the past few years have seen the publication of some excellent books providing an introduction to the field as a whole, as well as books covering other specific topics in cryptography.

functions. While I have made an effort to introduce all the necessary background material as needed, the reader will find things much more easy going if they have encountered this background material previously.

The current book is divided into three sections:

- *Part I — Setting the Stage.* This part includes relevant background material, an overview of digital signatures, and definitions of security for signature schemes. Even readers with a firm background in cryptography should skim this part of the book since the definitions given here include "non-standard" ones such as security against known-/random-message attacks, and "strong" security for signature schemes.

- *Part II — Digital Signature Schemes without Random Oracles.* Parts II and III of the book cover constructions of digital signature schemes. Part II focuses on schemes that can be proven secure without resorting to the "random oracle" model. (A brief introduction to the random oracle model is provided in Chapter 6.) This part begins with the important theoretical result showing that signatures can be constructed from any one-way function (though a complete proof is given only for the case of one-way permutations). Next, constructions based on the RSA and strong RSA assumptions are presented. Finally, some more recent constructions of signature schemes from bilinear maps are shown.

  To my knowledge, Part II describes essentially all known classes of signature schemes that do not rely on the random oracle model.

- *Part III — Digital Signature Schemes in the Random Oracle Model.* The signature schemes considered in Part II are, generally speaking, considered too inefficient for practical use. Instead, more efficient schemes with proofs of security in the random oracle model are used. Following a brief introduction to the random oracle model (along with a discussion of its pros and cons), we discuss the two main approaches used in constructing signatures in this setting: building signatures from identification schemes, and designing signatures using trapdoor permutations (or variants thereof) and the "hash-and-sign" approach.

Unfortunately omitted in this work is any discussion of signature schemes based on specific, "non number-theoretic" assumptions including those based on knapsacks, lattices, coding theory, or polynomial equations. I have also decided to focus only on "standard" signature schemes and not to cover any of the multitude of variants (e.g., undeniable, ring, group, homomorphic, . . . signature schemes) that are out there. From a basic theoretical perspective, however, this book is fairly comprehensive and will, I hope, serve as a useful primer for the more specialized literature.

Comments and Errata

I am always happy to receive feedback and constructive criticism enabling me to improve this book. I am also always grateful (though less happy) to hear about any errors or omissions. Please email any comments to jkatz@cs.umd.edu with "Digital Signatures Book" in the subject line.

Acknowledgments

It gives me great pleasure to acknowledge the unwavering support of my wife, Jill, during the time I wrote this book. I would also like to thank Yehuda Lindell and Bob Stern for allowing me to adapt some of the text from [72] for inclusion here. Finally, I would like to thank Susan Lagerstrom-Fife for her patience and encouragement (prodding?) during the course of this project.

Portions of this book were written during my sabbatical year at IBM. I am grateful to Tal Rabin and all the members of the crypto research group at IBM for being such wonderful hosts.

My work on this book was supported in part by the National Science Foundation under grants #0447075, #0627306, and #0716651. Any opinions, findings, conclusions, or recommendations expressed in this book are my own, and do not necessarily reflect the views of the National Science Foundation.

College Park, MD

Jonathan Katz
March 2010

Contents

Part II Digital Signature Schemes without Random Oracles

Part I
Setting the Stage

Chapter 1
Digital Signatures: Background and Definitions

1.1 Digital Signature Schemes: A Quick Introduction

Loosely speaking, a *digital signature scheme* offers a cryptographic analogue of handwritten signatures that, in fact, provides much stronger security guarantees. Digital signatures serve as a powerful tool and are now accepted as legally binding in many countries; they can be used for certifying contracts or notarizing documents, for authentication of individuals or corporations, and as components of more complex protocols. Digital signatures also enable the secure distribution and transmission of public keys and thus, in a very real sense, serve as the foundation for all of public-key cryptography.

A digital signature scheme is typically used by a *signer* and a set of potential *verifiers*. (Our discussion here will be relatively informal; we defer formal definitions until later.) The signer begins by running some *key-generation algorithm* to produce a pair of keys (pk, sk), where pk will be called the signer's *public key* — for reasons that will become obvious in a moment — and sk is the signer's *private key* (sometimes also referred to as its *secret key*). The signer then publicizes its public key, and we will assume that any potential verifier is in possession of (or can obtain) an authentic copy of the public key pk associated with the signer. We will not focus on the exact details of how a signer disseminates its public key; for concreteness, one can imagine that there is a public directory linking signers to their public keys, and this directory is administered in such a way that it is infeasible for someone to register a public key in someone else's name. We stress, however, that in general there will be many signers, each with their own public key, and so any potential verifier must know not only the set of valid public keys, but also which of these public keys belongs to the signer whose signature he is interested in verifying.

Once a signer has established a public key pk as discussed above, digital signature schemes allow the signer to "certify" (or "sign") a message in such a way that any other party who knows pk can verify that the message originated from the signer and has not been modified in any way. In more detail, for any message m (that we view simply as a bit-string) the signer can apply a *signing algorithm* to m using

J. Katz, *Digital Signatures*, DOI 10.1007/978-0-387-27712-7_1,
© Springer Science+Business Media, LLC 2010

its private key sk; this results in a signature σ that can be verified by anyone who knows pk using the corresponding *verification algorithm*.

It will be useful at this point to consider a prototypical usage of a digital signature scheme: Consider a software company that wants to issue software patches/updates in an authenticated manner; that is, when the company needs to release a software patch it should be possible for any of its clients to recognize that the patch is authentic, and a malicious third party should never be able to fool a client into accepting a patch that was not actually released by the company. To do this, the company can generate a public key pk along with a private key sk, and then distribute pk in some reliable manner to its clients (perhaps bundling the public key along with the initial distribution of the software). When releasing a patch m, the company can then compute a digital signature σ on m using its private key sk, and post (m, σ) on its webpage. Each client can verify the authenticity of m before downloading it by checking that σ is a legitimate signature on m with respect to the public key pk.

A malicious party might try to issue a fake patch by spoofing the company's webpage and posting (m', σ'), where m' represents a patch that was never released by the company. This m' might be a modified version of some previous patch m, or it might be completely new and unrelated to previous patches. If the signature scheme is "secure" (in a sense we will define more carefully soon), then when the client attempts to verify σ' it will find that this is an *invalid* signature on m' with respect to pk, and will therefore *reject* the signature. Note that in this application it is crucial that the client reject even if the forged patch m' is modified only slightly from a genuine patch m.

The above is not just a theoretical application of digital signatures, but one that is used extensively today. (E.g., Microsoft uses exactly this approach when issuing updates to its Windows operating system.)

The assumption that parties are able to obtain a legitimate copy of the signer's public key implies that the signer is able to transmit at least one message (namely, pk itself) in a reliable and authenticated manner. Given this, one may wonder why signature schemes are needed at all! The point is that reliable distribution of pk *is* a difficult task, but using a signature scheme means that this need only be carried out *once*, after which an unlimited number of messages can subsequently be sent reliably. Furthermore, signature schemes themselves are used to ensure the reliable distribution of *other* public keys as part of a public-key infrastructure (PKI).

1.1.1 Properties of Digital Signatures

We have just seen that digital signatures provide a means of authenticating messages sent over a public channel. Signature schemes provide stronger properties as well, and we elucidate these properties via a comparison with *message authentication codes*, the symmetric-key analogue of digital signatures.

An instance of a message authentication code is defined by a secret key s shared between a (single) sender and a (single) receiver. The sender can certify a message

m by applying a message authentication algorithm to m using the shared key s; this results in a "tag" t. The receiver, given m and t, can verify authenticity of m using a corresponding verification procedure along with the same key s. As with digital signatures, the security guarantee provided by a message authentication code is that no malicious third party, who does *not* know s, can forge a valid-looking tag t' on any message m' not explicitly authenticated by the sender. (We refer the reader to [72] for a much more in-depth discussion of message authentication codes.)

Thus, both message authentication codes and digital signature schemes can be used to ensure the integrity (or authenticity) of transmitted messages. One clear difference, however, is with respect to the initial key establishment phase. Digital signatures fall under the category of *public-key cryptography*, where a party (i.e., the signer in this case) need only distribute some key over a *public*, but authenticated, channel. With message authentication codes, on the other hand, the sender must share a key over a *secret* and authenticated channel. Signature schemes also have a decided advantage when a sender wants to issue the same message to multiple recipients. When using a digital signature scheme, this would be done by distributing a single public key and computing a single signature that can be verified by any potential recipient; in contrast, with a message authentication code the sender would have to establish a separate secret key with each possible receiver, and would have to compute a separate tag (with respect to the appropriate shared key) for each recipient as well.

A *qualitative* advantage that digital signatures have as compared to message authentication codes is that signatures are *publicly verifiable*. This means that if a receiver verifies the signature on a given message as being legitimate, then it is assured that all other parties who receive this signed message will also verify it as legitimate. This feature is not achieved by message authentication codes where a signer shares a separate key with each receiver: in such a setting a malicious sender might compute a correct tag with respect to receiver A's shared key but an incorrect tag with respect to a different user B's shared key. In this case, A knows that he received an authentic message from the sender but has no guarantee that other recipients will agree.

Public verifiability implies that signatures are *transferable*: a signature σ on a message m by a particular signer S can be shown to a third party, who can then verify herself that σ is a legitimate signature on m with respect to S's public key (here, we assume this third party also knows S's public key). By making a copy of the signature, this third party can then show the signature to another party and convince *them* that S authenticated m, and so on. Transferability and public verifiability are essential for the application of digital signatures to certificates and public-key infrastructures.

Digital signature schemes also provide the very important property of *non-repudiation*. That is — assuming a signer S widely publicizes his public key in the first place — once S signs a message he cannot later deny having done so. This aspect of digital signatures is crucial for situations where a recipient needs to prove to a third party (say, a judge) that a signer did indeed "certify" a particular message (e.g., a contract): assuming S's public key is known to the judge, or is otherwise publicly available, a valid signature on a message is enough to convince the judge that S indeed signed this message. Message authentication codes simply cannot provide this functionality. To see this, say users S and R share a key s_{SR}, and S sends a message m to R along with a (valid) MAC tag t computed using s_{SR}. Since the judge does *not* know s_{SR} (indeed, this key is kept secret by S and R), there is no way for the judge to determine whether t is a valid tag or not. If R were to reveal the key s_{SR} to the judge, there would still be no way for the judge to know whether this is the "actual" key that S and R shared, or whether it is some "fake" key manufactured, after the fact, by R. Even if were to assume that the judge is given the actual key s_{SR}, and can somehow be convinced of this fact, this still would not provide non-repudiation because there is no way for R to prove that it was S who generated t — the very fact that message authentication codes are *symmetric* (so that anything S can do, R can also do) implies that R could have generated t on its own, and so there is no way for the judge to distinguish between the actions of the two parties.

Because they are non-repudiable and publicly verifiable, digital signatures are used frequently to sign contracts, notarize documents, etc. over the Internet, and have been given legal validity in many countries.

Of course, a drawback of digital signatures as compared to message authentication codes is that the latter are roughly 2–3 orders of magnitude more efficient than the former. For this reason, in situations where public verifiability, transferability, and/or non-repudiation are not needed, and the sender will communicate primarily with a single recipient (with whom it is able to share a secret key), message authentication codes are preferred. We remark that there may also be settings where non-repudiation and transferability are specifically *not* desired: say, when a signer S wants a *particular* recipient to be assured that S certified a message, but does not want this recipient to be able to prove this fact to other parties. (This is sometimes referred to as the property of *deniability*.) In such a case, a message authentication code (or some more complicated cryptographic primitive) would have to be used.

1.2 Computational Security

One further difference between message authentication codes and digital signature schemes is that there exist message authentication codes that are *unconditionally* secure when a bounded number of messages are authenticated. Security of digital signature schemes, on the other hand, is inherently computational (even if we bound the number of messages being signed). Specifically, *no signature scheme can be secure against an all-powerful adversary* (an "all-powerful adversary" is one with unlimited computational power or, equivalently, unlimited time). Indeed, consider an adversary that, given a signer's public key pk and a message m, tries all possible values of σ until it finds one for which $\mathsf{Vrfy}_{pk}(m, \sigma) = 1$. (The adversary knows pk, and can therefore compute $\mathsf{Vrfy}_{pk}(\cdot, \cdot)$ on inputs of his choice.) Now, at least one σ satisfying $\mathsf{Vrfy}_{pk}(m, \sigma) = 1$ *exists*, since it must be possible for the legitimate signer to generate a valid signature on m. But then the adversary as described will eventually find such a σ and succeed in forging a signature.

A second observation is that *no signature scheme can be perfectly secure even against a very "weak" adversary.* Illustration of this point is even simpler than before: consider an adversary who simply chooses σ *at random.* Clearly, there is a non-zero probability that the value thus chosen will satisfy $\mathsf{Vrfy}_{pk}(m, \sigma) = 1$ (again, using the fact that at least one such σ satisfying this condition exists). This adversary does not even require knowledge of the signer's public key.

Does the above suggest that secure signature schemes are impossible to construct? Not at all. Thankfully, hope is not lost if one is willing to relax the security requirements and consider a *computational* notion of security rather than a *perfect* one. That is, instead of requiring (as above) that it be *impossible* for *any* adversary to forge a signature, we instead demand that it be impossible *except with "small" probability* for any *efficient* (i.e., computationally-bounded) adversary to forge a signature; note in particular that this rules out the two attacks sketched above. In this book, we formalize these notions in the standard way (see [72] for further discussion), and for completeness we briefly review the details now.

1.2.1 Computational Notions of Security

In moving to the computational setting, we introduce a *security parameter* $k \in \mathbb{N}$ that is used to parameterize both the adversary as well as the signature scheme itself; this security parameter can be viewed as quantifying the level of security obtained by a particular instance of the scheme (though this is not quite formally true). In a bit more detail, we view the signer as selecting a security parameter k when generating keys for the scheme; the security parameter will be passed as input to the key-generation algorithm and the length of the public and private keys will depend on k. Following the standard convention in theoretical cryptography, we equate "efficient adversaries" with algorithms running in probabilistic polynomial time (where the running time is measured as a function of k). Since the adversary will be limited to running in polynomial time it is only fair to require that all algorithms executed by the honest parties (e.g., signing, verification) should run in polynomial time as well. We abbreviate "probabilistic, polynomial time" as PPT.

An event occurs with "small probability" if the probability that the event occurs is *negligible* in k, where this is defined formally as follows:

Definition 1.1. A function $\varepsilon : \mathbb{N} \to [0, 1]$ is **negligible** if for all $c \geq 0$ there exists $k_c \geq 0$ such that $\varepsilon(k) < 1/k^c$ for all $k > k_c$.

Rephrased, then, a signature scheme satisfying a computational notion of security will have the property that every probabilistic polynomial-time adversary succeeds in forging a signature (with respect to this scheme) with only negligible probability. In other words, fix some adversary A running in polynomial time and let $\varepsilon_A(k)$ denote the probability that A forges a valid signature with respect to some signature scheme. (We will specify this experiment more carefully when we introduce formal security definitions.) Then the scheme is secure if $\varepsilon_A(k)$ is negligible in k, implying that as the signer increases the value of k, the "success" probability of the adversary A decreases rapidly. It should be stressed that this says nothing about the "absolute" security of the scheme for particular values of k; all guarantees are given only with respect to the asymptotic performance of the scheme. In some sense, then, a larger value of k results in a "more secure" scheme in practice; formally, though, a scheme is either secure or insecure without reference to any particular value of k.

This is best illustrated with an example.

Example 1.1. Imagine a secure signature scheme where the private key is a uniformly random string of length k, and consider the naive adversary A who performs a brute-force search for the private key for k^5 steps. (We assume the adversary can identify a private key that matches a given public key.) The probability that A finds the correct private key is $k^5/2^k$. For $k = 30$ the adversary runs for $2.4 \cdot 10^7$ steps and finds the private key with probability $\approx 1/50$; this would probably not be considered an acceptable level of security in practice (although the scheme itself — viewed asymptotically — is still "secure"). By "dialing up" the security parameter to $k = 60$ we already obtain a reasonable security guarantee: even if A runs for $7.7 \cdot 10^8$ steps, it finds the private key only with probability $6.7 \cdot 10^{-10}$.

The example above illustrates the importance of a *concrete* security analysis — that is, an analysis that quantifies the maximum success probability of any adversary running for a specific amount of time, and attacking a scheme using a specific value of the security parameter — in addition to the *asymptotic* one used in formal definitions of security. As noted already, a scheme proven secure using an asymptotic security analysis guarantees only that the probability that any polynomial-time adversary "breaks" the scheme becomes negligibly small as the security parameter is increased; it does not say what value of the security parameter to use *in practice* to ensure a particular level of security against an adversary running in a particular amount of time. This issue will be revisited in Chapter 7, where the concrete security analysis of some practical schemes will be considered.

1.2.2 Notation

In understanding the security of signature schemes, we will be interested in analyzing *probabilistic* experiments; we introduce some notation (following [61]) that will provide useful shorthand for doing so. Let A be a probabilistic Turing machine. Then $A(x;r)$ denotes the output[1] of A when A is run on input x and (sufficiently-long) random tape r. In case A takes multiple inputs $x_1,\ldots,x_n$ (that, of course, can always be encoded as a single input), then $A(x_1,\ldots,x_n;r)$ denotes the output of A when run on these inputs and random tape r.

---

[1] All A considered here will halt on all inputs and random tapes, so the value $A(x;r)$ is always well-defined.

In writing a multi-stage experiment, the notation $y := A(x;r)$ simply means that variable y is assigned the value $A(x;r)$. If S is a finite set, then $y \leftarrow S$ denotes assigning to y an element uniformly chosen from S. The notation $y_1, y_2 \leftarrow S$ is taken to mean that y_1 and y_2 are assigned elements uniformly and *independently* chosen from S. Generalizing this notation, $y \leftarrow A(x)$ refers to the experiment in which a random tape r (of the appropriate length) is selected uniformly at random, and then y is assigned the value $A(x;r)$; thus, if A uses a random tape of length (at most) ℓ on inputs of length $|x|$, the notation "$y \leftarrow A(x)$" is shorthand for:

$$ r \leftarrow \{0,1\}^{\ell}; y := A(x;r). $$

When A is *deterministic* $y := A(x)$ is equivalent to $y \leftarrow A(x)$, though we will use the former notation when we wish to emphasize that A is deterministic.

The probability of a particular event E following execution of some experiment expt is written as $\Pr[\text{expt} : E]$. Everything to the left of the colon represents the experiment itself (whose components are executed in order, from left to right), and the event of interest is written to the right of the colon. An event may be expressed as a predicate, where the event is said to occur if the predicate is true. As usual, $\wedge$ is the "logical and" operation, $\vee$ represents "logical or", $\{0,1\}^k$ denotes the set of binary strings of length k, and if E is an event then $\overline{E}$ denotes the complement of E (i.e., the event that E does *not* occur).

1.3 Defining Signature Schemes

Precise definitions are crucial if we are to understand the security guaranteed by any particular construction, and are essential before we can even hope to have rigorous proofs of security for the schemes we will develop. For completeness, we begin with a purely *functional* definition describing the basic functionality that any signature scheme should achieve; this is followed by a number of different *security* definitions detailing various levels of resilience a signature scheme might provide. In the remainder of the chapter, we focus on techniques for *amplifying* the security of a given signature scheme; i.e., we show how to take a scheme that achieves a relatively weak notion of security and adapt it so as to obtain a new scheme that realizes a stronger notion of security. Such techniques have proven to be very useful in the design and development of secure signature schemes, and we will see many examples of these techniques in the constructions described in later chapters.

Definition 1.2. A **signature scheme** consists of three probabilistic, polynomial-time algorithms (Gen, Sign, Vrfy) along with an associated message space $\mathcal{M} = \{M_k\}$ such that:

- The randomized **key-generation algorithm** Gen takes as input the security parameter k (in unary). It outputs a pair of keys (pk, sk) where pk is called the **public key** or the **verification key**, and sk is called the **private key**, the **secret**

key, or the **signing key**. We assume the security parameter k is implicit in both pk and sk.

- For security parameter k, the (possibly randomized) **signing algorithm** Sign takes as input a secret key sk and a message $m \in M_k$. It outputs a signature σ. We write this as $\sigma \leftarrow \text{Sign}_{sk}(m)$. We assume that if $m \notin M_k$, the signature algorithm outputs a distinguished symbol $\perp$.
- For security parameter k, the deterministic **verification algorithm** Vrfy takes as input a public key pk, a message $m \in M_k$, and a (purported) signature σ. It outputs a single bit b, with $b = 1$ signifying "accept" and $b = 0$ signifying "reject." We write this as $b := \text{Vrfy}_{pk}(m, \sigma)$. We assume that if $m \notin M_k$, the verification algorithm rejects.

When a given public key is understood from the context, we say a message/signature pair (m, σ) is **valid** if $\text{Vrfy}_{pk}(m, \sigma) = 1$. We require that for all k, all (pk, sk) output by $\text{Gen}(1^k)$, all $m \in M_k$, and all σ output by $\text{Sign}_{sk}(m)$, the message/signature pair (m, σ) is valid.

It should be clear that the above definition exactly formalizes the intuitive notion described earlier. Specifically, a signature scheme is used in the following way. One party S, who acts as the *signer*, runs $\text{Gen}(1^n)$ to obtain keys (pk, sk). When S wants to transmit a message m, it computes the signature $\sigma \leftarrow \text{Sign}_{sk}(m)$ and sends (m, σ). Upon receipt of (m, σ), a receiver who knows pk can verify the authenticity of m by checking whether $\text{Vrfy}_{pk}(m, \sigma) \stackrel{?}{=} 1$. This establishes both that S sent m, and also that m was not modified in transit. It is worth emphasizing, however, that successful verification does not prove anything about *when* m was sent, and thus so-called *replay attacks* are possible. We return to this point later.

Relaxations of the Definition

We briefly discuss some relaxations of Definition 1.2.

The correctness requirement in the definition can be relaxed to allow for a negligible probability of error. We ignore this issue for the most part in our presentation, though we caution the reader that some of the schemes we present do indeed have negligible error probability.

Another variation of the above definition allows the possibility of *randomized* verification algorithms, allowing a negligible probability of error when a correctly-generated signature is verified. Few schemes in the literature use randomized verification, and we will not see any examples in this book.

Instead of requiring that the message space be fixed *a priori* (for each value of k), it is sometimes convenient to allow the set of legal messages to depend on the public key generated by Gen. When, on occasion, this will be the case for schemes described in this book, it will be evident from the context and so we will not explicitly mention it. Note that for any practical application the message space should consist of all bit-strings (perhaps of some bounded length).

It is possible for the signing algorithm to be *stateful* (rather than *stateless* as defined above): formally, in this case some *state s* would be initialized to NULL by Gen, and the signing algorithm would take as input the current state *s* in addition to the message and the secret key, and would output an updated value *s'* for the state in addition to the signature. (We stress that the state is not needed in order to verify the signature. Also, the adversary is assumed unable to view the state of the signer unless otherwise specified.) For the most part, it is viewed as undesirable in practice for a signature scheme to be stateful; the security of some signature schemes, however, crucially depends on the ability of the signing algorithm to maintain such state. (We will see one such example in Chapter 3.) Unless otherwise specified, we always assume stateless signature schemes.

1.4 Motivating the Definitions of Security

The basic security guarantee offered by a signature scheme is, roughly speaking, that no efficient adversary should be able to "forge" a valid message/signature pair with respect to a public key *pk* generated by an honest user, even after the adversary "interacts" with this user and thereby obtains legitimate signatures on multiple messages, generated using the private key *sk* associated with *pk*. As outlined previously, we will equate "efficient" with the notion of (probabilistic) polynomial time, and will allow the adversary to possibly succeed in outputting a valid forgery with some negligibly-small probability (recall that both the adversary's running time and its probability of forging a signature are measured as a function of the security parameter k). To formally specify the desired notion of security, then, it remains to define two things more precisely: first, what it means to "forge" a valid message/signature pair; second, what exactly "interacting" with the legitimate user entails. These two components may be specified independently, leading to numerous possible definitions; we explore some of these possibilities here.

Let us first, still somewhat informally, discuss some plausible interpretations of what it means to "forge" a valid message/signature pair. One immediate observation is that if the legitimate signer himself generates the signatures $\sigma_1, \ldots, \sigma_\ell$ on the messages $m_1, \ldots, m_\ell$, respectively, then the adversary cannot be prevented from *replaying* a valid message/signature pair (m_i, σ_i). Clearly, then, such "replay" attacks should not count as forgeries and we will not try to prevent attacks of this sort in our definitions.[2] Given this, a natural approach is to say that an adversary succeeds in constructing a forgery if it generates a valid message/signature pair (m, σ) such that m is *not* one the messages previously signed by the legitimate signer (i.e., $m \notin \{m_1, \ldots, m_\ell\}$); we will refer to this a forgery of this type as an *existential forgery* and call a scheme in which such forgeries are infeasible *existentially unforgeable*.

---

[2] When using a signature scheme as a component of some larger protocol, however, it is certainly important to recognize that such replay attacks can occur and — if needed — protect against such attacks by ensuring that all messages requiring a signature are distinct (by using timestamps or nonces, for example).

We stress that it is not required that m be "meaningful" in any way; indeed, the notion of what is "meaningful" is application-dependent and it therefore does not make much sense to incorporate any semantics (or, equivalently, any notion of "meaningfulness") in a definition of security for signature schemes. This has an added benefit: any signature scheme that is existentially unforgeable can be used in *any* application that requires access to the functionality provided by signatures, without having to worry about whether the semantics of the application are compatible with any semantics assumed by the signature scheme itself.

A stronger notion termed *strong* existential unforgeability has recently been introduced, and is required for some applications of digital signature schemes. Here, the adversary is said to produce a forgery even if it outputs a *new* (valid) signature on a *previously signed* message (this is in addition to the notion of existential unforgeability defined above); formally, a forgery occurs whenever the adversary outputs a valid message/signature pair $(m, \sigma) \notin \{(m_1, \sigma_1), \ldots, (m_\ell, \sigma_\ell)\}$ (where, as before, σ_i represents a signature on m_i that was obtained from the legitimate signer). Notice that this gives a strict generalization of existential unforgeability.

Having discussed the notion of forgery, we turn next to consideration of the possible ways in which an adversary might interact with the legitimate signer. We focus specifically on two features of this interaction: (1) *how many* signatures the legitimate signer generates, and (2) *what sort of control* the adversary has over the messages that are signed. With regard to the first question, we will distinguish only between the case where a *single* signature is generated and the case in which an *unbounded* number of signatures might be generated. Signature schemes intended for the setting when only a single signature is produced are called *one-time* signatures. Besides the historical reasons for considering this, relatively weak, definition (in particular, the first provably-secure signature scheme was a one-time signature scheme), one-time signatures also serve as useful building blocks of standard signature schemes (as we will see later). One-time signature schemes are also useful in their own right for certain applications.

The second question (namely, the amount of control the adversary has over the messages that are signed) is more interesting. Here, we describe informally three different scenarios we will consider, and provide some justification as to why each represents a natural class of adversarial behavior in certain settings. We stress, though, that our ultimate goal is to achieve security in the final (strongest) scenario; schemes secure within the context of the other two models, however, will serve as useful "stepping stones" toward this goal (see especially Sections 1.7.1 and 1.7.2).

Random-message attack: The first scenario we will consider may be viewed as modeling the case wherein the adversary has no control whatsoever over the messages that are signed. As an example, one can imagine here that the adversary merely observes signatures that are produced on messages provided by *other* (honest) parties. We formalize this by considering an adversary who is given signatures on a sequence of *random* messages. Although this may seem a poor approximation to the motivating scenario we have just discussed, a little thought shows that no other reasonable option is available without introducing some (*a priori* unjustified) assumption regarding the distribution of messages that are signed. Furthermore, there are scenarios in which the messages that are signed are, indeed, random and this then provides an exact model of an adversary's attack; a good example is when a user is authenticated by computing a signature on a random challenge sent by a verifier.

Known-message attack: The adversary is now assumed to have some limited control over what messages are signed, or there may be some definite pattern in the types of messages likely to be signed (and so a random-message attack is no longer the most appropriate model). We will formalize this attack scenario by granting the adversary control over the messages that are signed, subject to the restriction that the adversary must specify these messages *in advance* and, in particular, *independently* of the signer's public key as well as any of the subsequent signatures the adversary observes. One can view this as modeling a "worst-case" choice of messages to be signed, as long as these messages are not directly under the adversary's control (and so cannot be changed once the public key is fixed).

(Adaptive) chosen-message attack: This model is the strongest possible, in that it grants the adversary complete control over what messages are signed. Specifically, the adversary can not only choose these messages *after* it sees the public key, but can also select messages based on previous signatures (on previously chosen messages) that the adversary observes; hence the term "adaptive." (We will often omit this term, however, since we do not consider any non-adaptive variant of chosen-message attacks.)

One might object that a realistic adversary would never have *complete* control over what is signed, since surely the legitimate signer will refuse to sign certain messages (e.g., "I am a crook"). As discussed earlier in the context of forgery, however, it makes little sense to impose any semantics on the message space in a definition of security. Furthermore, there may well be cases (e.g., a signer acting as a notary) when an adversary might have a significant amount of control over what is signed. Finally, a major advantage of such a strong attack model — as in the case of our strong definition of what constitutes a forgery — is that a scheme proven secure against such a strong attack will certainly be secure for any desired application (without having to consider whether the needs of the application "match" the guarantees provided by the security definition).

With the above in mind, we will introduce a number of formal definitions in the following section corresponding to various combinations of a notion of forgery and an attack model. Specifically, we will define the notions of (1) existential unforgeability for one-time signature schemes; (2) existential unforgeability under random-message attack; (3) existential unforgeability under known-message attack; and (4) (strong) existential unforgeability under adaptive chosen-message attack. We introduce the first definition, as we have mentioned, for its historical significance as well as for its usefulness in constructing schemes satisfying stronger definitions of security, as we will see in a later chapter. Similarly, the second and third definitions will be used in Sections 1.7.1 and 1.7.2 to construct schemes satisfying the fourth (strongest) notion. The last definition we consider is the strongest possible (at least with regard to the possibilities discussed above), and has become the *de facto* notion of security that any "good" signature scheme should satisfy. One may argue that this definition is "overkill" with respect to what is needed in practice, although we have argued the contrary above. Even if this were true, however, the definition is surely sufficient for any application of signature schemes in practice, and thus there is a substantial benefit to constructing schemes satisfying this strong notion of security.

1.5 Formal Definitions of Security

We state our definitions in a slightly different order than that discussed in the previous paragraph.

1.5.1 Security against Random-Message Attacks

We begin by defining security against random-message attacks. Here, recall, the adversary is given no control over what messages are signed; instead, these messages are chosen randomly from the message space. For completeness, we define both existential unforgeability and *strong* existential unforgeability.

Definition 1.3. Signature scheme $(\mathsf{Gen}, \mathsf{Sign}, \mathsf{Vrfy})$ is **existentially unforgeable under a random-message attack** if for all polynomials $\ell(\cdot)$ and all probabilistic, polynomial-time adversaries A, the success probability of A in the following experiment is negligible (as a function of k):

1. A sequence of $\ell = \ell(k)$ messages $m_1, \ldots, m_\ell$ are chosen uniformly at random from the message space M_k.
2. The key-generation algorithm $\mathsf{Gen}(1^k)$ is run to obtain a pair of keys (pk, sk).
3. Signatures $\sigma_1 \leftarrow \mathsf{Sign}_{sk}(m_1), \ldots, \sigma_\ell \leftarrow \mathsf{Sign}_{sk}(m_\ell)$ are computed.
4. A is given pk and $\{(m_i, \sigma_i)\}_{i=1}^{\ell}$, and outputs (m, σ).
5. A **succeeds** if (1) $\mathsf{Vrfy}_{pk}(m, \sigma) = 1$ and (2) $m \notin \{m_1, \ldots, m_\ell\}$.

The scheme is **strongly unforgeable under a random-message attack** if for all polynomials $\ell(\cdot)$ and all probabilistic, polynomial-time adversaries A, the success probability of A is negligible (as a function of k) in an experiment defined as above except for the last condition, now changed to:

5′. A **succeeds** if (1) $\mathsf{Vrfy}_{pk}(m, \sigma) = 1$ and (2) $(m, \sigma) \notin \{(m_1, \sigma_1), \ldots, (m_\ell, \sigma_\ell)\}$.

As a technical point, if the message space depends on the public key then the order of the above experiments is changed so that the public and secret keys are generated *first*, and then the messages are chosen at random from the message space thus defined. (The remainder of the experiment is unchanged.)

Just to get comfortable with the notation, note that we can express the definition above in the following compact way. Signature scheme $(\mathsf{Gen}, \mathsf{Sign}, \mathsf{Vrfy})$ is existentially unforgeable under a random-message attack if for all polynomials $\ell(\cdot)$ and all probabilistic, polynomial-time adversaries A it is the case that $\varepsilon_A(k)$ is negligible, where:

$$\varepsilon_A(k) \stackrel{\text{def}}{=} \Pr\left[\begin{array}{c} \{m_i\}_{i=1}^{\ell} \leftarrow M_k;\ (pk, sk) \leftarrow \mathsf{Gen}(1^k);\\ \forall i \in [\ell]: \sigma_i \leftarrow \mathsf{Sign}_{sk}(m_i);\\ (m, \sigma) \leftarrow A\left(pk, \{(m_i, \sigma_i)\}_{i=1}^{\ell}\right) \end{array} : \begin{array}{c} \mathsf{Vrfy}_{pk}(m, \sigma) = 1 \wedge\\ m \notin \{m_1, \ldots, m_\ell\} \end{array}\right].$$

We will continue to write out experiments in full when we present definitions, but will switch to the more compact notation when writing proofs of security.

It is sometimes also useful to consider a very weak adversary who is allowed to obtain a signature on only a *single* message.

Definition 1.4. Signature scheme $(\mathsf{Gen}, \mathsf{Sign}, \mathsf{Vrfy})$ is **existentially unforgeable under a one-time random-message attack** if for all probabilistic, polynomial-time adversaries A, the success probability of A in the following experiment is negligible (as a function of k):

1. A message m_1 is chosen uniformly at random from the message space M_k.

2. The key-generation algorithm $\mathsf{Gen}(1^k)$ is run to obtain a pair of keys (pk, sk).

3. Signature $\sigma_1 \leftarrow \mathsf{Sign}_{sk}(m_1)$ is computed.

4. A is given pk and (m_1, σ_1), and outputs (m, σ).

5. A **succeeds** if (1) $\mathsf{Vrfy}_{pk}(m, \sigma) = 1$ and (2) $m \neq m_1$.

The scheme is **strongly unforgeable under a one-time random-message attack** if for all probabilistic, polynomial-time adversaries A, the success probability of A is negligible (as a function of k) in an experiment defined as above except for the last condition, now changed to:

5′. A **succeeds** if (1) $\mathsf{Vrfy}_{pk}(m, \sigma) = 1$ and (2) $(m, \sigma) \neq (m_1, \sigma_1)$.

One can, by extension, also consider ℓ-time attacks for any polynomial ℓ. Note, however, the distinction between the resulting notion of security and Definition 1.3: in the former case, ℓ is fixed *a priori* and the signature scheme is allowed to depend on ℓ, whereas in the case of Definition 1.3 the scheme is fixed and then security is required to hold for any (polynomial) ℓ.

1.5.2 Security against Known-Message Attacks

We now define security against known-message attacks. Here, recall, the adversary is able to choose what messages are signed, but it must make this choice in advance of seeing the public key. Again, for completeness, we include a definition of strong unforgeability also.

Definition 1.5. Signature scheme $(\mathsf{Gen}, \mathsf{Sign}, \mathsf{Vrfy})$ is **existentially unforgeable under a known-message attack** if for all polynomials $\ell(\cdot)$ and all probabilistic, polynomial-time adversaries A, the success probability of A in the following experiment is negligible (as a function of k):

1. $A(1^k)$ outputs a sequence of $\ell = \ell(k)$ messages $m_1, \ldots, m_\ell \in M_k$.

2. The key-generation algorithm $\mathsf{Gen}(1^k)$ is run to obtain a pair of keys (pk, sk).

3. Signatures $\sigma_1 \leftarrow \mathsf{Sign}_{sk}(m_1), \ldots, \sigma_\ell \leftarrow \mathsf{Sign}_{sk}(m_\ell)$ are computed.

4. A is given pk and $\{\sigma_i\}_{i=1}^{\ell}$, and outputs (m, σ).

5. A **succeeds** if (1) $\mathsf{Vrfy}_{pk}(m, \sigma) = 1$ and (2) $m \notin \{m_1, \ldots, m_\ell\}$.

(We assume that A is a stateful algorithm, and in particular is allowed to maintain state between steps 1 and 4.)

The scheme is **strongly unforgeable under a known-message attack** if for all polynomials $\ell(\cdot)$ and all probabilistic, polynomial-time adversaries A, the success probability of A is negligible (as a function of k) in an experiment defined as above except for the last condition, now changed to:

5′. A **succeeds** if (1) $\mathsf{Vrfy}_{pk}(m, \sigma) = 1$ and (2) $(m, \sigma) \notin \{(m_1, \sigma_1), \ldots, (m_\ell, \sigma_\ell)\}$.

A subtle point arises, again, in case the message space depends on the public key; in that case, it becomes difficult (or impossible) to formally define a notion of security against known-message attacks. In almost any "natural" signature scheme, however — and certainly any scheme used in practice — there is some set M_k' of messages that is always contained in the message space, regardless of the public key. (For example, M_k' might consist of all bit-strings of some given length, such that these strings can be embedded in the message space regardless of the public key.) We can then restrict the adversary to outputting messages in M_k'. Another possibility occurs when the public key can be split naturally into two components, only the first of which determines the message space; in this case, the experiment can be modified so the adversary is given only the first component of the public key before being asked to generate the messages to be signed. In the rest of the book, we will implicitly make assumptions of this sort when dealing with security against known-message attacks.

We can, of course, also define existential unforgeability (and strong unforgeability) under a *one-time* known-message attack. Since a definition is exactly analogous to Definition 1.4, we do not give such a definition here.

1.5.3 Security against Adaptive Chosen-Message Attacks

We now define the strongest attack model, in which the adversary can adaptively choose messages to be signed, depending both on the public key as well as on any previous signatures it has obtained. To capture this formally, we will provide the adversary *oracle access* to a so-called *signing oracle* $\mathsf{Sign}_{sk}(\cdot)$. This oracle should be thought of as a "black box" that computes signatures under the fixed secret key sk: the adversary can submit any input m to this oracle, and will receive in return the signature $\sigma \leftarrow \mathsf{Sign}_{sk}(m)$. An important point is that the signing key sk used by the oracle is exactly the one generated along with the public key pk in the experiment; that is, the adversary is obtaining signatures that are valid with respect to public key it is "attacking." It is to be stressed that the signing oracle is not meant to represent any physical device to which the adversary has access in the real world. Rather, this oracle is simply a convenient means to model the adversary's interaction with the signer, under the assumption that the adversary can convince the signer to sign any messages of the adversary's choice (as justified in Section 1.4).

Definition 1.6. Signature scheme $(\mathsf{Gen}, \mathsf{Sign}, \mathsf{Vrfy})$ is **existentially unforgeable under a chosen-message attack** if for all probabilistic, polynomial-time adversaries A, the success probability of A in the following experiment is negligible (as a function of k):

1. The key-generation algorithm $\mathsf{Gen}(1^k)$ is run to obtain a pair of keys (pk, sk).

2. A is given pk and allowed to interact with a signing oracle $\mathsf{Sign}_{sk}(\cdot)$, requesting signatures on as many messages as it likes. (We denote this by $A^{\mathsf{Sign}_{sk}(\cdot)}(pk)$.) Let M denote the set of messages queried to the signing oracle by A.

3. Eventually, A outputs (m, σ).

4. A **succeeds** if (1) $\mathsf{Vrfy}_{pk}(m, \sigma) = 1$ and (2) $m \notin M$.

The scheme is **strongly unforgeable under a chosen-message attack** if for all probabilistic, polynomial-time adversaries A, the success probability of A in the following experiment is negligible (as a function of k):

1. The key-generation algorithm $\mathsf{Gen}(1^k)$ is run to obtain a pair of keys (pk, sk).

2. A is given pk and allowed to interact with a signing oracle $\mathsf{Sign}_{sk}(\cdot)$, requesting signatures on as many messages as it likes. Let $Q = \{(m_i, \sigma_i)\}$ where m_i denotes the ith query made by A to the signing oracle, and σ_i is the ith response.

3. Eventually, A outputs (m, σ).

4. A **succeeds** if (1) $\mathsf{Vrfy}_{pk}(m, \sigma) = 1$ and (2) $(m, \sigma) \notin Q$.

(In both definitions, we assume that A is a stateful algorithm.)

The attack considered in the above definition is often called an *adaptive* chosen-message attack in the literature, but since we do not consider any "non-adaptive" variant we often drop the extra qualifier.

Since we will refer to it often, we also define the corresponding notion of security for one-time attacks.

Definition 1.7. Signature scheme $(\mathsf{Gen}, \mathsf{Sign}, \mathsf{Vrfy})$ is **existentially unforgeable under a one-time chosen-message attack** if for all probabilistic, polynomial-time adversaries A, the success probability of A in the following experiment is negligible (as a function of k):

1. The key-generation algorithm $\text{Gen}(1^k)$ is run to obtain a pair of keys (pk, sk).

2. A is given pk and allowed to request a signature on a single message m_1. In return it is given $\sigma_1 \leftarrow \text{Sign}_{sk}(m_1)$.

3. A outputs (m, σ).

4. A **succeeds** if (1) $\text{Vrfy}_{pk}(m, \sigma) = 1$ and (2) $m \neq m_1$.

The scheme is **strongly unforgeable under a one-time chosen-message attack** if for all probabilistic, polynomial-time adversaries A, the success probability of A in the following experiment is negligible (as a function of k):

1. The key-generation algorithm $\text{Gen}(1^k)$ is run to obtain a pair of keys (pk, sk).

2. A is given pk and allowed to request a signature on a single message m_1. In return it is given $\sigma_1 \leftarrow \text{Sign}_{sk}(m_1)$.

3. A outputs (m, σ).

4. A **succeeds** if (1) $\text{Vrfy}_{pk}(m, \sigma) = 1$ and (2) $(m, \sigma) \neq (m_1, \sigma_1)$.

(As usual, both definitions assume that A is a stateful algorithm.)

1.5.3.1 Terminology

By default, when we talk about "security" of a digital signature scheme we mean existential unforgeability (with respect to the appropriate attack model); we often also write "unforgeability" in place of "existential unforgeability". For brevity, we occasionally refer to security against random-message attacks, known-message attacks, and chosen-message attacks as RMA-security, KMA-security, and CMA-security, respectively.

By way of terminology, we say that an adversary A "forges a signature on a new message" or "outputs a forgery" whenever A outputs a message/signature pair (m, σ) such that $\text{Vrfy}_{pk}(m, \sigma) = 1$ and A was not previously given any signature on m. We say that A "outputs a strong forgery" whenever A outputs a message/signature pair (m, σ) such that $\text{Vrfy}_{pk}(m, \sigma) = 1$ and A was not previously given the signature σ on the message m. Note that whenever A outputs a forgery then it also outputs a strong forgery.

1.6 Relations Between the Notions

The definitions in the previous section lead to a strict hierarchy of security guarantees (assuming that there exist secure signature schemes to begin with). That is:

- There exist signature schemes that are existentially unforgeable under a *one-time* chosen-message attack that are not existentially unforgeable under a chosen-message attack (indeed, it is not hard to see that the scheme we show in Section 3.1 has exactly this property), and similarly for random-message attacks and known-message attacks.

- There exist signature schemes that are existentially unforgeable under a random-message attack but that are not existentially unforgeable under a known-message attack. Similarly, there are signature schemes that are existentially unforgeable under a known-message attack but that are not existentially unforgeable under a chosen-message attack. In both cases, constructing a convoluted scheme proving these assertions is relatively straightforward but we are unaware of any "natural" schemes with these properties. (We stress, however, that a number of natural schemes can be proven secure with respect to a weaker notion of security while no proof is known that they satisfy a stronger notion of security.)
- There exist signature schemes that are existentially unforgeable under a chosen-message attack, but that are not *strongly* unforgeable (even if the adversary obtains a signature on only a single random message).

The above observations motivate the importance of choosing a scheme satisfying the appropriate definition of security for the application at hand; one cannot simply "assume" that a scheme satisfying a weak definition automatically satisfies stronger definitions as well. On the other hand, if a particular application requires a signature scheme only satisfying a weaker notion of security, then one can hope to improve efficiency by using a scheme satisfying exactly that notion (and nothing stronger).

1.7 Achieving CMA-Security from Weaker Primitives

As discussed at the end of the previous section, the weaker notions of security we have defined — security against random-message attacks and security against known-message attacks — may provide meaningful guarantees in certain restricted settings. More interesting, perhaps, is their usefulness is constructing schemes satisfying our strongest definition: security against chosen-message attacks. In this section, we explore this possibility and show how schemes secure with respect to a weaker notion can be converted to schemes secure within the strongest attack model. The constructions we show are not only interesting from a theoretical point of view but have practical significance as well: numerous schemes rely on the ideas used in the following constructions, and the conversions themselves are relatively efficient (decreasing the efficiency of the original, weaker scheme by roughly a factor of two).

1.7.1 CMA-Security from RMA-security

We begin by showing how to construct a CMA-secure scheme based on any RMA-secure one. The basic idea is to 'split' each message m being signed into two random pieces m_L, m_R subject to the constraint that $m_L \oplus m_R = m$. Then m_L and m_R are signed using two independent instances of the RMA-secure scheme. (To prevent an adversary from mixing-and-matching pieces from two different signed messages, a random nonce is chosen and signed along with each piece.) The details follow.

Construction 1.1: CMA-security from RMA-security

Let $\Pi = (\text{Gen}, \text{Sign}, \text{Vrfy})$ be a signature scheme for message of length $k + q(k)$. Construct the signature scheme $\Pi^* = (\text{Gen}^*, \text{Sign}^*, \text{Vrfy}^*)$ for messages of length $q = q(k)$ as follows:

Key generation: Algorithm $\text{Gen}^*(1^k)$ is defined as follows:

- Run two (independent) invocations of $\text{Gen}(1^k)$ to obtain the keys (pk_L, sk_L), (pk_R, sk_R). (We use "L" and "R" to stand for "left" and "right".)
- The public key is $pk^* = (pk_L, pk_R)$ and the secret key is $sk^* = (sk_L, sk_R)$.

Signature generation: Algorithm $\text{Sign}^*_{sk^*}(m)$ is defined as follows:

- Parse sk^* as sk_L, sk_R.
- Choose $r \leftarrow \{0,1\}^k$ and $m_L \leftarrow \{0,1\}^q$. Set $m_R := m \oplus m_L$.
- Compute $\sigma_L \leftarrow \text{Sign}_{sk_L}(r\|m_L)$ and $\sigma_R \leftarrow \text{Sign}_{sk_R}(r\|m_R)$, where "$\|$" denotes concatenation.
- Output the signature $\sigma^* = (r, m_L, m_R, \sigma_L, \sigma_R)$.

Signature verification: Algorithm $\text{Vrfy}^*_{pk}(m, \sigma^*)$ is defined as follows: Parse σ^* as $(r, m_L, m_R, \sigma_L, \sigma_R)$ and pk as (pk_L, pk_R). Then, output 1 iff $m \stackrel{?}{=} m_L \oplus m_R$ and both

$$\text{Vrfy}_{pk_L}(r\|m_L, \sigma_L) \stackrel{?}{=} 1$$

and

$$\text{Vrfy}_{pk_R}(r\|m_R, \sigma_R) \stackrel{?}{=} 1.$$

It is easy to see that the scheme above is correct. (It is not strictly necessary to include m_R in the signature output by Sign^*, but it makes the description of the scheme a bit more transparent.) The following theorem shows that the above construction achieves the desired level of security:

Theorem 1.1. *If $\Pi = (\text{Gen}, \text{Sign}, \text{Vrfy})$ is existentially unforgeable (resp., strongly unforgeable) under a random-message attack, then $\Pi^* = (\text{Gen}^*, \text{Sign}^*, \text{Vrfy}^*)$ given by Construction 1.1 is existentially unforgeable (resp., strongly unforgeable) under an adaptive chosen-message attack.*

Proof. We focus on the case of strong unforgeability, but the claim regarding existential unforgeability follows by examination of the proof. The intuition is simple: in scheme Π^*, the "messages" signed by the underlying signature scheme Π (namely, $r\|m_L$ and $r\|m_R$) are always individually uniform, regardless of the adversary's choice of m. Of course, the *joint* distribution of $r\|m_L$ and $r\|m_R$ depends on m, but it is their individual distributions we are interested in since $r\|m_L$ and $r\|m_R$ are each signed using a different secret key. This suggests that security of Π against a random-message attack is sufficient to prove security of Π^* against a chosen-message attack. The only hitch is that if a nonce value r is ever used twice, then signature forgery becomes trivial. For example, if an adversary obtains the signature $(r, m_L, m_R, \sigma_L, \sigma_R)$ on message m and signature $(r, m'_L, m'_R, \sigma'_L, \sigma'_R)$ on message m', then this means that σ_L is a valid signature on $r\|m_L$ with respect to pk_L, and σ'_R is a valid signature on $r\|m'_R$ with respect to pk_R. Thus, the adversary can output

$$(r, m_L, m'_R, \sigma_L, \sigma'_R),$$

a valid signature on the message $m_L \oplus m'_R$ (and this message is likely not equal to either of m or m'). Fortunately, it is not hard to show that the probability of using any nonce twice is small.

Turning to the formal proof, let A^* be a PPT adversary attacking Π^* and denote by $(m, \sigma^*) \leftarrow \mathsf{Expt}_{A^*, \Pi^*}(1^k)$ the experiment

$$(pk^*, sk^*) \leftarrow \mathsf{Gen}^*(1^k); (m, \sigma^*) \leftarrow (A^*)^{\mathsf{Sign}^*_{sk}(\cdot)}(pk^*).$$

Let Forge be the event that $\mathsf{Vrfy}^*_{pk^*}(m, \sigma^*) = 1$ and $(m, \sigma^*) \notin Q$, where Q is as in Definition 1.6. Define

$$\mathsf{Succ}_{A^*, \Pi^*}(k) \overset{\text{def}}{=} \Pr\left[(m, \sigma^*) \leftarrow \mathsf{Expt}_{A^*, \Pi^*}(1^k) : \mathsf{Forge}\right],$$

and note that this is exactly the success probability of A^* as defined in Definition 1.6. Thus, our goal is to show that $\mathsf{Succ}_{A^*, \Pi^*}(k)$ is negligible.

Let $\ell = \ell(k)$ denote the maximum number of queries made by A^* to its signing oracle on security parameter k, and assume without loss of generality that A^* always makes exactly this many queries; note that ℓ must be polynomial since A^* runs in polynomial time. In a given execution of $\mathsf{Expt}_{A^*, \Pi^*}(1^k)$, let m_i denote the ith message submitted by A^* to its signing oracle, and let $\sigma^*_i = (r_i, m_{L,i}, m_{R,i}, \sigma_{L,i}, \sigma_{R,i})$ denote the ith signature received. Denote by $\sigma^* = (r^*, m^*_L, m^*_R, \sigma^*_L, \sigma^*_R)$ the components of the signature output by A^*.

Now define the following events:

- Let Repeat denote the event that two signatures obtained by A^* from its signing oracle use the same value for the nonce r, that is, that $r_i = r_j$ for some $i \neq j$.
- Let $Q_L = \{(r_i\|m_{L,i}, \sigma_{L,i})\}$ denote the set of "left" message/signature pairs. Let Forge_L denote the event that $\mathsf{Vrfy}_{pk_L}(r^*\|m^*_L, \sigma^*_L) = 1$ and $(r^*\|m^*_L, \sigma^*_L) \notin Q_L$.
- Similarly, let $Q_R = \{(r_i\|m_{R,i}, \sigma_{R,i})\}$ be the set of "right" message/signature pairs, and let Forge_R be the event that $\mathsf{Vrfy}_{pk_R}(r^*\|m^*_R, \sigma^*_R) = 1$ and $(r^*\|m^*_R, \sigma^*_R) \notin Q_R$.

We claim that whenever Forge occurs, at least one of Repeat, Forge_L, or Forge_R occurs. To see this, assume Forge occurs and Repeat does not occur. Since Forge occurs, we know that $\mathsf{Vrfy}_{pk_L}(r^*\|m_L^*,\sigma_L^*)=1$ and $\mathsf{Vrfy}_{pk_R}(r^*\|m_R^*,\sigma_R^*)=1$. Since Repeat does not occur, we know that $r^*=r_i$ for at most one value of i. There are two cases to consider:

Case 1: r^* is not equal to r_i for any value of i. In this case, we clearly have $r^*\|m_L^*\neq r_i\|m_{L,i}$ for all i and so, in particular, $(r^*\|m_L^*,\sigma_L^*)\notin Q_L$. This means that Forge_L occurs. (By a symmetric argument, Forge_R occurs in this case as well.)

Case 2: $r^*=r_i$ for some (unique) i. If both $(m_L^*,\sigma_L^*)=(m_{L,i},\sigma_{L,i})$ and $(m_R^*,\sigma_R^*)=(m_{R,i},\sigma_{R,i})$, then we have $m=m_i$ and $\sigma^*=\sigma_i^*$ in contradiction to the fact that $(m,\sigma^*)\notin Q$ (since Forge occurred). Thus, it must be the case that either $(m_L^*,\sigma_L^*)\neq(m_{L,i},\sigma_{L,i})$ (in which case Forge_L occurs) or $(m_R^*,\sigma_R^*)\neq(m_{R,i},\sigma_{R,i})$ (in which case Forge_R occurs).

We conclude that

$$\Pr[\mathsf{Expt}_{A^*,\Pi^*}(1^k):\mathsf{Forge}]\leq \tag{1.1}$$
$$\Pr[\mathsf{Expt}_{A^*,\Pi^*}(1^k):\mathsf{Forge}_L]+\Pr[\mathsf{Expt}_{A^*,\Pi^*}(1^k):\mathsf{Forge}_R]$$
$$+\Pr[\mathsf{Expt}_{A^*,\Pi^*}(1^k):\mathsf{Repeat}].$$

We show that each of the terms on the right-hand side is negligible, thus completing the proof of the theorem. We deal with the term that is easiest to analyze first.

Claim. $\Pr[\mathsf{Expt}_{A^*,\Pi^*}(1^k):\mathsf{Repeat}]$ is negligible.

Proof. The claim follows easily from a "birthday problem" calculation. (See [72, Appendix A.4] for further information.) Specifically, We have ℓ nonces $r_1,\ldots,r_\ell$ chosen uniformly from the set $\{0,1\}^k$ of size 2^k. The probability that two of these values are equal is at most $\ell^2/2^{k+1}$, which is negligible in k.

Claim. $\Pr[\mathsf{Expt}_{A^*,\Pi^*}(1^k):\mathsf{Forge}_L]$ is negligible.

Proof. To prove this claim, we reduce to the strong unforgeability of Π. Consider the following PPT adversary A, using A^* as a subroutine and attacking Π in a random-message attack:

> **Algorithm A:**
> The algorithm is given a public key pk, generated using $\mathsf{Gen}(1^k)$, along with ℓ signatures $\sigma_1,\ldots,\sigma_\ell$ on the random messages $m_1',\ldots,m_\ell'\in\{0,1\}^{k+q}$.
>
> - Set $pk_L:=pk$.
> - Run $\mathsf{Gen}(1^k)$ to obtain keys (pk_R,sk_R).
> - Set $pk^*:=(pk_L,pk_R)$ and run $A^*(pk^*)$.
> - When A^* requests a signature on the ith message m_i, do:
>   1. Let r_i be the first k bits of m_i', and let $m_{L,i}$ be the last q bits of m_i'.
>   2. Set $\sigma_{L,i}:=\sigma_i$.

3. Set $m_{R,i} := m_i \oplus m_{L,i}$.
4. Compute $\sigma_{R,i} \leftarrow \text{Sign}_{sk_R}(r_i \| m_R)$.
5. Return the signature $(r_i, m_{L,i}, m_{R,i}, \sigma_{L,i}, \sigma_{R,i})$ to A^*.

- When A^* outputs $(m, \sigma^* = (r^*, m_L^*, m_R^*, \sigma_L^*, \sigma_R^*))$, then output $(r^* \| m_L^*, \sigma_L^*)$.

Observe first that A provides a perfect simulation for A^*; that is, the view of A^* when run by A (and when pk is generated by $\text{Gen}(1^k)$) is *identically distributed* to the view of A^* in $\text{Expt}_{A^*, \Pi^*}(1^k)$. It is easy to see that the public key pk^* is identically distributed in both experiments. As for the answers to the signing queries of A^*, note that these, too, are distributed identically in both cases; here, we rely on the fact that the m_i' are chosen uniformly (and independently) at random.

To conclude the proof, we merely observe that A outputs a strong forgery whenever Forge_L occurs. So, the success probability of A attacking Π in the sense of Definition 1.3 (i.e., with respect to a random-message attack) is exactly equal to $\Pr[\text{Expt}_{A^*, \Pi^*}(1^k) : \text{Forge}_L]$. Since Π is assumed to be strongly unforgeable under a random-message attack, this must be negligible.

An exactly analogous argument shows that $\Pr[\text{Expt}_{A^*, \Pi^*}(1^k) : \text{Forge}_R]$ is negligible. Since each term on the right-hand side of Equation (1.1) is negligible, we conclude that the success probability of A^* in attacking Π^* under a chosen-message attack is negligible. This completes the proof of the theorem.

1.7.2 CMA-Security from KMA-Security

In this section we show how to construct a signature scheme that is secure against chosen-message attacks from any scheme secure under a known-message attack. Note that since any KMA-secure scheme is also RMA-secure, we could just as well use the construction from the previous section. Nevertheless, the construction described here offers an alternate approach that is sometimes more efficient when applied to particular schemes.

The key idea here is to use a level of indirection. Let Π be a KMA-secure scheme, and let Π' be a KMA-secure *one-time* signature scheme. We construct a CMA-secure scheme Π^* as follows: the public key pk is generated using Π. To sign a message m, the signer generates a fresh public key pk' using Π'; signs pk' with respect to pk; and signs m with respect to pk'. Note that this "indirection" removes any dependence between any "message" and the public key with respect to which it is signed: m is independent of pk', and pk' is independent of pk. Details follow.

We stress that a *fresh* key pair (pk', sk') is generated each time a new message is signed. It is easy to see that the above scheme is correct. We now prove that it realizes the desired level of security.

Construction 1.2: CMA-security from KMA-security

Let $\Pi' = (\mathsf{Gen}', \mathsf{Sign}', \mathsf{Vrfy}')$ be a signature scheme in which the public keys have length $q = q(k)$, and let $\Pi = (\mathsf{Gen}, \mathsf{Sign}, \mathsf{Vrfy})$ be a signature scheme in which the message space M_k includes all bit-strings of length $q(k)$. Construct the signature scheme $\Pi^* = (\mathsf{Gen}^*, \mathsf{Sign}^*, \mathsf{Vrfy}^*)$, whose message space is exactly the message space of Π', as follows:

Key generation: $\mathsf{Gen}^*(1^k)$ simply runs $\mathsf{Gen}(1^k)$ to generate keys (pk, sk). These are the public and secret keys, respectively.

Signature generation: Algorithm $\mathsf{Sign}^*_{sk}(m)$ is defined as follows:

1. Run $\mathsf{Gen}'(1^k)$ to generate keys (pk', sk').
2. Sign pk' using sk: i.e., compute $\sigma \leftarrow \mathsf{Sign}_{sk}(pk')$.
3. Sign m using sk'; i.e., compute $\sigma' \leftarrow \mathsf{Sign}'_{sk'}(m)$.
4. Output the signature $\sigma^* = (pk', \sigma, \sigma')$.

Signature verification: Algorithm $\mathsf{Vrfy}^*_{pk}(m, \sigma^*)$ is defined as follows: Parse σ^* as (pk', σ, σ'). Then output 1 if and only if σ is a valid signature on pk' (with respect to pk) and σ' is a valid signature on m (with respect to pk'); i.e., output 1 if and only if both

$$\mathsf{Vrfy}_{pk}(pk', \sigma) \overset{?}{=} 1$$

and

$$\mathsf{Vrfy}'_{pk'}(m, \sigma') \overset{?}{=} 1.$$

Theorem 1.2. *If $\Pi = (\mathsf{Gen}, \mathsf{Sign}, \mathsf{Vrfy})$ is existentially unforgeable (resp., strongly unforgeable) under a known-message attack and $\Pi' = (\mathsf{Gen}', \mathsf{Sign}', \mathsf{Vrfy}')$ is unforgeable (resp., strongly unforgeable) under a one-time known-message attack, then $\Pi^* = (\mathsf{Gen}^*, \mathsf{Sign}^*, \mathsf{Vrfy}^*)$ is existentially unforgeable (resp., strongly unforgeable) under an adaptive chosen-message attack.*

Proof. We prove the theorem for the case of existential unforgeability, but a proof for the case of strong unforgeability can be derived by making suitable modifications. The intuition here is again rather straightforward. Assume the adversary obtains a sequence of signatures $(pk'_1, \sigma_1, \sigma'_1), \dots$ on messages $m_1, \dots$ chosen adaptively by the adversary. A key observation is that if an adversary forges a signature (pk', σ, σ') on a new message m, then it must be the case that either $pk' = pk'_i$ for some i, or not. (We assume here for simplicity $pk'_i \neq pk'_j$ for all $i \neq j$; as we will see, this assumption is not needed in the formal proof.) If $pk' = pk'_i$, then the adversary has effectively forged a signature σ' on new message m with respect to scheme Π' and the public key pk'_i; otherwise, the adversary has effectively forged a signature σ on the new "message" pk' with respect to scheme Π and public key pk. Furthermore, we may note that all "messages" being signed are chosen *independently* of the corresponding public key: in particular, the keys $\{pk_i'\}$ that are signed using PK are chosen by the signer (who generates them independently of each other as well as independently of the key PK), and each message m_i — although chosen by the adversary — is chosen *before* the adversary knows the value of the key pk_i' that will be used to sign m_i. For this reason, security of Π, Π' under a *known* message attack is sufficient.[3] Moreover, it suffices for Π' to be unforgeable under a one-time known-message attack since every key pk_i' is used to sign only a single message m_i.

We now proceed with the formal proof. Given a PPT adversary A^* attacking the signature scheme Π^*, denote by $(m, pk', \sigma, \sigma') \leftarrow \mathsf{Expt}_{A^*, \Pi^*}(1^k)$ the experiment:

$$(pk, sk) \leftarrow \mathsf{Gen}^*(1^k); (m, pk', \sigma, \sigma') \leftarrow (A^*)^{\mathsf{Sign}_{sk}^*(\cdot)}(pk).$$

Let m_i denote the ith message submitted by A^* to its signing oracle, and let $\{(pk_i', \sigma_i, \sigma_i')\}$ denote the ith signature received in return. Let Forge be the event that $\mathsf{Vrfy}_{pk}^*(m, pk', \sigma, \sigma') = 1$ and $m \notin \{m_i\}$, and define

$$\mathsf{Succ}_{A^*, \Pi^*}(k) \stackrel{\mathrm{def}}{=} \Pr\left[(m, pk', \sigma, \sigma') \leftarrow \mathsf{Expt}_{A^*, \Pi^*}(1^k) : \mathsf{Forge}\right];$$

this is exactly the success probability of A^* in attacking scheme Π^* in the sense of Definition 1.6. Our goal is to show that $\mathsf{Succ}_{A^*, \Pi^*}(k)$ is negligible.

We will now condition on whether or not A^* re-uses one of the keys pk_i' in his attempted forgery. That is, let Reuse be the event that $pk' = pk_i'$ for some i, and then define

$$\mathsf{Succ}_{A^*, \Pi^*}^{\overline{\mathsf{Resue}}}(k) \stackrel{\mathrm{def}}{=} \Pr\left[(m, pk', \sigma, \sigma') \leftarrow \mathsf{Expt}_{A^*, \Pi^*}(1^k) : \mathsf{Forge} \wedge \overline{\mathsf{Reuse}}\right]$$

$$\mathsf{Succ}_{A^*, \Pi^*}^{\mathsf{Reuse}}(k) \stackrel{\mathrm{def}}{=} \Pr\left[(m, pk', \sigma, \sigma') \leftarrow \mathsf{Expt}_{A^*, \Pi^*}(1^k) : \mathsf{Forge} \wedge \mathsf{Reuse}\right].$$

Of course, we have:

$$\mathsf{Succ}_{A^*, \Pi^*}^*(k) = \mathsf{Succ}_{A^*, \Pi^*}^{\overline{\mathsf{Resue}}}(k) + \mathsf{Succ}_{A^*, \Pi^*}^{\mathsf{Reuse}}(k).$$

We show that each of the terms on the right-hand side is negligible. We assume without loss of generality in what follows that A^* always requests signatures on exactly $\ell = \ell(k)$ messages for some polynomial ℓ.

Claim. $\mathsf{Succ}_{A^*, \Pi^*}^{\overline{\mathsf{Resue}}}(k)$ is negligible.

Proof. We construct a PPT adversary A, using A^* as a subroutine, that attacks Π in a known-message attack and succeeds with probability exactly $\mathsf{Succ}_{A^*, \Pi^*}^{\overline{\mathsf{Resue}}}(k)$. Security of Π thus yields the claim. The algorithm A is defined as follows:

---

[3] Observe further that if the public keys $\{pk_i'\}$ output by Gen' are random strings, then it is enough for Π to be unforgeable under a *random* message attack.

Algorithm A:

- Run $\mathsf{Gen}'(1^k)$ a total of ℓ times to obtain keys $\{(pk_i', sk_i')\}_{i=1}^{\ell}$.
- Output $pk_1', \ldots, pk_\ell'$. Receive in return a public key pk along with the signatures $\{\sigma_i\}_{i=1}^{\ell}$. (Note: each σ_i is a valid signature on pk_i' with respect to pk.)
- Run $A^*(pk)$. When A^* requests a signature on the ith message m_i, do:
  1. Compute $\sigma_i' \leftarrow \mathsf{Sign}'_{sk_i'}(m_i)$.
  2. Return the signature $(pk_i', \sigma_i, \sigma_i')$ to A^*.
- When A^* outputs $(m, \sigma^* = (pk', \sigma, \sigma'))$, output (pk', σ).

It is immediate that the view of A^* in the above experiment is identical to its view in $\mathsf{Expt}_{A^*,\Pi^*}(1^k)$; hence, the event $\mathsf{Forge} \wedge \overline{\mathsf{Reuse}}$ occurs in the above execution of A with probability exactly $\mathsf{Succ}_{A^*,\Pi^*}^{\overline{\mathsf{Resue}}}(k)$. Since Reuse does not occur, we have $pk' \notin \{pk_i'\}$; since Forge occurs, it must be the case that σ is a valid signature on pk' (with respect to the public key pk). Hence, we conclude that A outputs a forgery on a new message with probability $\mathsf{Succ}_{A^*,\Pi^*}(k)$. The assumed security of Π under a known-message attack implies that this is negligible.

Claim. $\mathsf{Succ}_{A^*,\Pi^*}^{\mathsf{Resue}}(k)$ is negligible.

Proof. We now construct a PPT adversary A' attacking Π' in a one-time known-message attack and having success probability at least $\mathsf{Succ}_{A^*,\Pi^*}^{\mathsf{Resue}}(k)/\ell$. Since ℓ is polynomial and Π' is, by assumption, existentially unforgeable under a one-time known-message attack, the claim follows.

Algorithm A':

- Compute $(pk, sk) \leftarrow \mathsf{Gen}(1^k)$ and choose a random index $i^* \leftarrow \{1, \ldots, \ell\}$.
- Run $A^*(pk)$, answering its ith signature query for a message m_i as follows:

  Case 1: $i \neq i^*$.
  1. Run $\mathsf{Gen}'(1^k)$ to generate keys (pk_i', sk_i').
  2. Compute $\sigma_i \leftarrow \mathsf{Sign}_{sk}(pk_i')$ and $\sigma_i' \leftarrow \mathsf{Sign}_{sk_i'}(m_i)$.
  3. Return the signature $(pk_i', \sigma_i, \sigma_i')$ to A^*.

  Case 2: $i = i^*$.
  1. Output m_{i^*} and receive a public key pk_{i^*}' and a signature σ_{i^*}'.
  2. Compute $\sigma_{i^*} \leftarrow \mathsf{Sign}_{sk}(pk_{i^*}')$.
  3. Return the signature $(pk_{i^*}', \sigma_{i^*}, \sigma_{i^*}')$ to A^*.

- When A^* outputs $(m, \sigma^* = (pk', \sigma, \sigma'))$, if $pk' = pk_{i^*}'$ then output (m, σ') (else output nothing).

That is, A' chooses a random index i^* and answers all signature queries of A^* normally *except* for the i^*th query; A' can do this since it knows the "master" secret key sk corresponding to pk. For the i^*th query for a signature, on the message m_{i^*}, the adversary A' outputs m_{i^*} to its *own* "oracle" and receives in return a randomly generated public key pk'_{i^*} (generated using $\mathsf{Gen}'(1^k)$) along with a signature σ'_{i^*} on m_{i^*}. It then computes the signature σ_{i^*} on pk'_{i^*} on its own (using sk) and returns $(pk'_{il}, \sigma_{i^*}, \sigma'_{i^*})$ to A^*.

It is fairly easy to see that the view of A^* in the above experiment is identical to its view in $\mathsf{Expt}_{A^*, \Pi^*}(1^k)$, and therefore the probability that $\mathsf{Forge} \wedge \mathsf{Reuse}$ occurs in the above experiment is exactly $\mathsf{Succ}^{\mathsf{Resue}}_{A^*, \Pi^*}(k)$. When Reuse occurs, there is at least one index i for which $pk' = pk'_i$; since the distribution of i^* is uniform given the view of A^*, we have that $pk' = pk'_{i^*}$ with probability at least $1/\ell$. Given that Forge occurs it must be the case that $m \neq m_{i^*}$ and $\mathsf{Vrfy}'_{pk'_{i^*}}(m, \sigma') = 1$, and so A' outputs a valid forgery in this case. In summary, A' outputs a valid forgery with probability at least $\mathsf{Succ}^{\mathsf{Resue}}_{A^*, \Pi^*}(k)/\ell$. The claim follows.

The preceding two claims complete the proof.

Better efficiency using pre-computation. Besides being useful for constructing a CMA-secure scheme from a KMA-secure one, Construction 1.2 can also be used to improve the efficiency of signature generation using *pre-computation* that does not depend on the message being signed. (Signature schemes where pre-computation can be applied to improve efficiency are sometimes called "on-line/off-line" schemes.) Specifically, steps 1 and 2 of signature generation are independent of the message being signed; hence they can be carried out in advance and the results cached. When the message m to be signed is known, all that is needed is to compute a signature on m using the underlying one-time signature scheme. As we will see in Section 8.2.3, one-time signature schemes with very efficient signing can be constructed based on a variety of number-theoretic assumptions.

1.8 From Unforgeability to Strong Unforgeability

In the preceding two sections we amplified the security of a signature scheme in terms of the *attack* the scheme was able to withstand. Here we amplify the *notion of unforgeability*, showing how to convert a scheme that is unforgeable (under a chosen-message attack) to one that is *strongly* unforgeable. We use as a building block a signature scheme that is strongly unforgeable under a one-time chosen-message attack; we will see later on in the book that such schemes are relatively easy to construct.

Theorem 1.3. *If $\Pi = (\mathsf{Gen}, \mathsf{Sign}, \mathsf{Vrfy})$ is existentially unforgeable under an adaptive chosen-message attack and $\Pi' = (\mathsf{Gen}', \mathsf{Sign}', \mathsf{Vrfy}')$ is strongly unforgeable un-*

Construction 1.3: Strong unforgeability from unforgeability

Let $\Pi = (\mathsf{Gen}, \mathsf{Sign}, \mathsf{Vrfy})$ and $\Pi' = (\mathsf{Gen}', \mathsf{Sign}', \mathsf{Vrfy}')$ be signature schemes; for simplicity, assume they can each be used to sign messages of unbounded length (cf. the construction in the following section). Consider the following construction of signature scheme $\Pi^* = (\mathsf{Gen}^*, \mathsf{Sign}^*, \mathsf{Vrfy}^*)$:

Key generation: $\mathsf{Gen}^*(1^k)$ simply runs $\mathsf{Gen}(1^k)$ to generate keys (pk, sk). These are the public and secret keys, respectively.

Signature generation: Algorithm $\mathsf{Sign}^*_{sk}(m)$ is defined as follows:

- Run $\mathsf{Gen}'(1^k)$ to generate keys (pk', sk').
- Sign $pk' \| m$ using sk: i.e., compute $\sigma \leftarrow \mathsf{Sign}_{sk}(pk' \| m)$.
- Sign σ using sk'; i.e., compute $\sigma' \leftarrow \mathsf{Sign}'_{sk'}(\sigma)$.
- Output the signature $\sigma^* = (pk', \sigma, \sigma')$.

Signature verification: Algorithm $\mathsf{Vrfy}^*_{pk}(m, \sigma^*)$ is defined as follows: Parse σ^* as (pk', σ, σ'). Then output 1 if and only if σ is a valid signature on $pk' \| m$ (with respect to pk) and σ' is a valid signature on σ (with respect to pk'); i.e., output 1 if and only if

$$\mathsf{Vrfy}_{pk}(pk' \| m, \sigma) \overset{?}{=} 1 \quad \text{and} \quad \mathsf{Vrfy}'_{pk'}(\sigma, \sigma') \overset{?}{=} 1.$$

---

der a one-time chosen-message attack, then $\Pi^* = (\mathsf{Gen}^*, \mathsf{Sign}^*, \mathsf{Vrfy}^*)$ is strongly unforgeable under an adaptive chosen-message attack.

Proof. Intuition for strong unforgeability is straightforward. Consider a sequence of signatures $(pk'_1, \sigma_1, \sigma'_1), \ldots, (pk'_\ell, \sigma_\ell, \sigma'_\ell)$ obtained on messages $m_1, \ldots, m_\ell$, respectively. With overwhelming probability, each of the $\{pk'_i\}$ are distinct. Let $(m, (pk', \sigma, \sigma'))$ be a strong forgery output by an adversary. We may distinguish several cases:

Case 1: $pk' \notin \{pk'_i\}$. This clearly results in a forgery with respect to scheme Π.

Case 2: $pk' = pk'_i$ for some unique i. Note we must have $m = m_i$ or else we have a forgery with respect to Π. There are then two sub-cases:

- $\sigma \neq \sigma_i$. This gives a forgery with respect to scheme Π' (and public key pk'_i).
- $\sigma = \sigma_i$. This implies $\sigma' \neq \sigma'_i$ (else there is no strong forgery with respect to Π^*), but then this gives a *strong* forgery with respect to Π' (and public key pk'_i).

For a formal proof, let A^* be a PPT adversary attacking Π^*, and denote by $(m, pk', \sigma, \sigma') \leftarrow \mathsf{Expt}_{A^*, \Pi^*}(1^k)$ the experiment:

$$(pk, sk) \leftarrow \mathsf{Gen}^*(1^k); (m, pk', \sigma, \sigma') \leftarrow (A^*)^{\mathsf{Sign}^*_{sk}(\cdot)}(pk).$$

Let m_i denote the ith message submitted by A^* to its signing oracle, and let $\{(pk_i', \sigma_i, \sigma_i')\}$ denote the ith signature received in return. Let Forge be the event that $\mathsf{Vrfy}_{pk}^*(m, pk', \sigma, \sigma') = 1$ and $(m, \sigma^* = (pk', \sigma, \sigma')) \notin Q$, where Q is as in Definition 1.6. Define

$$\mathsf{Succ}_{A^*,\Pi^*}^*(k) \overset{\text{def}}{=} \Pr\left[(m, pk', \sigma, \sigma') \leftarrow \mathsf{Expt}_{A^*,\Pi^*}(1^k) : \mathsf{Forge} \right];$$

as usual, this is exactly the success probability of A^* in attacking scheme Π^* in the sense of strong unforgeability as defined in Definition 1.6. Our goal is to show that $\mathsf{Succ}_{A^*,\Pi^*}^*(k)$ is negligible.

As in the proof of Theorem 1.2, we will condition on whether or not A^* re-uses one of the keys pk_i' in his attempted forgery. Here, however, we define this event slightly differently: Reuse is now the event that $(m, pk') = (m_i, pk_i')$ for some i. Define

$$\mathsf{Succ}_{A^*,\Pi^*}^{\overline{\mathsf{Reuse}}}(k) \overset{\text{def}}{=} \Pr\left[(m, pk', \sigma, \sigma') \leftarrow \mathsf{Expt}_{A^*,\Pi^*}(1^k) : \mathsf{Forge} \wedge \overline{\mathsf{Reuse}} \right]$$

$$\mathsf{Succ}_{A^*,\Pi^*}^{\mathsf{Reuse}}(k) \overset{\text{def}}{=} \Pr\left[(m, pk', \sigma, \sigma') \leftarrow \mathsf{Expt}_{A^*,\Pi^*}(1^k) : \mathsf{Forge} \wedge \mathsf{Reuse} \right].$$

Of course,

$$\mathsf{Succ}_{A^*,\Pi^*}^*(k) = \mathsf{Succ}_{A^*,\Pi^*}^{\overline{\mathsf{Resue}}}(k) + \mathsf{Succ}_{A^*,\Pi^*}^{\mathsf{Resue}}(k).$$

We show that each of the terms on the right-hand side is negligible, thus proving the theorem.

Claim. $\mathsf{Succ}_{A^*,\Pi^*}^{\overline{\mathsf{Reuse}}}(k)$ is negligible.

The proof of this claim is almost identical to the proof of the first claim in the proof of Theorem 1.2, and is therefore omitted.

Claim. $\mathsf{Succ}_{A^*,\Pi^*}^{\mathsf{Reuse}}(k)$ is negligible.

Proof. The proof of this claim is very similar to the proof of the second claim in the proof of Theorem 1.2; because we deal with *strong* unforgeability here, we provide the details.

Assume without loss of generality that A^* always makes exactly $\ell = \ell(k)$ signature queries. We construct a PPT adversary A' attacking Π' in a one-time chosen-message attack and having success probability (in the sense of strong unforgeability) at least $\mathsf{Succ}_{A^*,\Pi^*}^{\mathsf{Resue}}(k)/\ell$. Since ℓ is polynomial and Π' is, by assumption, existentially unforgeable under a one-time known-message attack, the claim follows.

> **Algorithm A':**
> The algorithm is given a public key PK' (generated using $\mathsf{Gen}'(1^k)$) and is given access to a signing oracle $\mathsf{Sign}_{SK'}(\cdot)$ that it will query once.
>
> - Compute $(pk, sk) \leftarrow \mathsf{Gen}(1^k)$ and choose a random index $i^* \leftarrow \{1, \ldots, \ell\}$. Set $pk_{i^*}' := PK'$.

- Run $A^*(pk)$, answering its ith signature query for a message m_i as follows:

  Case 1: $i \neq i^*$.
  1. Run $\mathsf{Gen}'(1^k)$ to generate keys (pk_i', sk_i').
  2. Compute $\sigma_i \leftarrow \mathsf{Sign}_{sk}(pk_i' \| m_i)$ and $\sigma_i' \leftarrow \mathsf{Sign}_{sk_i'}(\sigma_i)$.
  3. Return the signature $(pk_i', \sigma_i, \sigma_i')$ to A^*.

  Case 2: $i = i^*$.
  1. Compute $\sigma_{i^*} \leftarrow \mathsf{Sign}_{sk}(pk_{i^*}' \| m_{i^*})$.
  2. Query σ_{i^*} to the signing oracle and receive in return a signature σ_{i^*}'.
  3. Return the signature $(pk_{i^*}', \sigma_{i^*}, \sigma_{i^*}')$ to A^*.

- When A^* outputs $(m, \sigma^* = (pk', \sigma, \sigma'))$, if $pk' = pk_{i^*}'$ then output (m, σ') (else output nothing).

As in the proof of the previous theorem, the view of A^* in the above experiment is identical to its view in $\mathsf{Expt}_{A^*, \Pi^*}(1^k)$, and therefore the probability that Forge $\wedge$ Reuse occurs in the above experiment is exactly $\mathsf{Succ}_{A^*, \Pi^*}^{\mathsf{Resue}}(k)$. When Reuse occurs, there is at least one index i for which $(m, pk') = (m_i, pk_i')$; since the distribution of i^* is uniform given the view of A^*, we have that $(m, pk') = (m_{i^*}, pk_{i^*}')$ with probability at least $1/\ell$. Assume this to be the case in what follows.

Since Forge occurs we know that $\mathsf{Vrfy}_{pk_{i^*}'}'(\sigma, \sigma') = 1$. Furthermore, we must have $(m, pk', \sigma, \sigma') \neq (m_{i^*}, pk_{i^*}', \sigma_{i^*}, \sigma_{i^*}')$. (Recall that here, Forge means that A^* output a *strong* forgery.) Since we are assuming $(m, pk') = (m_{i^*}, pk_{i^*}')$, this implies that $(\sigma, \sigma') \neq (\sigma_{i^*}, \sigma_{i^*}')$. But this means that A' outputs a strong forgery.

In summary, A' outputs a strong forgery with probability at least $\mathsf{Succ}_{A^*, \Pi^*}^{\mathsf{Resue}}(k)/\ell$. The claim, and thus the theorem, follows.

1.9 Extending the Message Length

We conclude this chapter by demonstrating how a signature scheme for k-bit messages can be extended to give a signature scheme for messages of arbitrary and variable length. (Technically we handle only messages shorter than $2^{k/4}$ bits, but since this bound is exponential it is not a serious restriction.) While we will see other means of accomplishing the same goal in the next chapter, the advantage of the present transformation is that it does not require any additional primitives.

Theorem 1.4. *If Π is existentially unforgeable (resp., strongly unforgeable) under an adaptive chosen-message attack, then Π^* is existentially unforgeable (resp., strongly unforgeable) under an adaptive chosen-message attack.*

Proof. We show that (except with negligible probability) a forgery with respect to Π^* implies a forgery with respect to Π. Turning this into a formal proof is left as an

---

Construction 1.4: From "short" messages to arbitrary length messages

---

Let $\Pi = (\mathrm{Gen}, \mathrm{Sign}, \mathrm{Vrfy})$ be a signature scheme for k-bit messages. Construct signature scheme $\Pi^* = (\mathrm{Gen}^*, \mathrm{Sign}^*, \mathrm{Vrfy}^*)$ for messages of length less than $2^{k/4}$ as follows:

Key generation: $\mathrm{Gen}^*(1^k)$ simply runs $\mathrm{Gen}(1^k)$ to generate keys (pk, sk). These are the public and secret keys, respectively.

Signature generation: Algorithm $\mathrm{Sign}^*_{sk}(m)$ is defined as follows:

- Let $L < 2^{k/4}$ be the length of m, and parse m into ℓ blocks $m_1, \ldots, m_\ell$, each of length $k/4$. (The final block is padded with 0s if necessary, though any such padding is *not* counted when determining the length L.)
- Choose a random identifier $r \leftarrow \{0,1\}^{k/4}$.
- For $i = 1, \ldots, \ell$, compute $\sigma_i \leftarrow \mathrm{Sign}_{sk}(r\|L\|i\|m_i)$, where L and i are uniquely encoded as strings of length $k/4$.
- Output the signature $\sigma := \langle r, \sigma_1, \ldots, \sigma_\ell \rangle$.

Signature verification: Algorithm $\mathrm{Vrfy}^*_{pk}(m, \sigma)$ is defined as follows:

- Let $L < 2^{k/4}$ be the length of m, and parse m into ℓ blocks $m_1, \ldots, m_\ell$, each of length $k/4$ (padding with 0s if necessary, though again not counting this padding when determining the length).
- Parse σ as $(r, \sigma_1, \ldots, \sigma_{\ell'})$.
- Output 1 if and only if $\ell' = \ell$ and $\mathrm{Vrfy}_{pk}(r\|L\|i\|m_i, \sigma_i) = 1$ for $1 \leq i \leq \ell$.

---

exercise. We deal with the case of existential unforgeability, but the case of strong unforgeability is essentially the same.

Let pk be the public key of the signer. Assume a total of $q = q(k)$ messages have been signed, and let $r^{(i)}$ be the identifier chosen by the signer when signing the ith message. Observe that the $\{r^{(i)}\}_{i=1}^q$ are all distinct, except with negligible probability. For the remainder of the proof we assume this to be the case, and show that (under this assumption) a forgery with respect to Π^* implies a forgery with respect to Π.

Consider some forgery (m, σ) with respect to Π^*. Let $L < 2^{k/4}$ be the length of m, and parse m into ℓ blocks $m_1, \ldots, m_\ell$, each of length $k/4$ (padding with 0s as usual). Parse σ as $(r, \sigma_1, \ldots, \sigma_\ell)$. There are two cases:

Case 1: $r \notin \{r^{(i)}\}_{i=1}^q$. Since (m, σ) is a valid forgery, we have in particular

$$\mathrm{Vrfy}_{pk}(r\|L\|1\|m_1, \sigma_1) = 1,$$

yet no (k-bit) message with prefix r was ever signed (with respect to Π). So $(r\|L\|1\|m_1, \sigma_1)$ is a forgery with respect to Π.

Case 2: $r = r^{(i)}$ **for some unique** i. (Uniqueness follows from out assumption that all the $\{r^{(i)}\}$ are distinct.) Let $m^{(i)}$ denote the ith message that was signed and let $L^{(i)}$ denote its length. If $L \neq L^{(i)}$ then $(r\|L\|1\|m_1, \sigma_1)$ is clearly a forgery with respect to Π, so assume $L = L^{(i)}$. Parse $m^{(i)}$ as $m_1^{(i)}, \ldots, m_\ell^{(i)}$, and let j be the first index where $m_j^{(i)} \neq m_j$ (there must be some such index, since $m^{(i)} \neq m$ but their lengths are the same). Then $(r\|L\|j\|m_j, \sigma_j)$ is a forgery with respect to Π. (All other blocks signed using Π differ in either the first or third components; the one previously signed block $r\|L\|j\|m_j^{(i)}$ differs from $r\|L\|j\|m_j$.) This completes the proof.

A similar proof applies also to the case of KMA-security:

Theorem 1.5. *If Π is existentially unforgeable (resp., strongly unforgeable) under a known-message attack, then Π^* is existentially unforgeable (resp., strongly unforgeable) under a known-message attack.*

1.10 Further Reading

The idea of rigorously defining a computational notion of security, and of proving security of a construction by reducing its security to a more basic assumption, is due to the pioneering work of Goldwasser and Micali [59] in the context of public-key encryption. For a good overview of the concept of "provable security" as well as a discussion about concrete vs. asymptotic security, see [72].

The basic idea of digital signature schemes was first suggested in the highly influential article by Diffie and Hellman [40] that initiated the study of public-key cryptography. In their work, Diffie and Hellman proposed both the concept as well as a generic instantiation; although their approach for constructing signature schemes turned out to be insecure (something that was not altogether obvious until definitions of security for signature schemes were formalized), their ideas served as the impetus for future work. Along with introducing the RSA cryptosystem (see Chapter 2), Rivest, Shamir, and Adleman [99] suggested its use for constructing a basic signature scheme. Other notable early work on digital signatures includes [97, 80, 110, 78, 44, 79].

One-time signatures were first considered by Lamport [76], and known-message attacks were studied by Goldwasser, Micali, and Yao [62] (albeit without completely formal definitions in either case). These papers also contain secure constructions meeting the proposed definitions. The definition of existential unforgeability under an adaptive chosen-message attack, which has become the "default" notion of security for digital signatures, was introduced in the seminal paper of Goldwasser, Micali, and Rivest [61]; their paper also contains a nice discussion about various possible definitions of security for signature schemes, including some that are not discussed here. Goldwasser, Micali, and Rivest also provide a construction satisfying their strongest definition. (Interestingly, the authors of [61] seem not to have recognized at the time that the construction of Section 1.7.2 could have been used to convert the KMA-secure scheme from [62] into a CMA-secure scheme.) The notion of *strong* unforgeability was popularized by the work of [7] (where an analogous notion was defined in the symmetric-key setting) and that of [2]. Further discussion about notions of security for digital signature schemes (including some additional notions not described here) can be found in the textbook by Goldreich [57].

Even, Goldreich, and Micali [45] proved Theorem 1.2 using essentially the same construction shown here, and the underlying ideas have been implicit in many constructions of digital signature schemes since then. Even, Goldreich, and Micali were also the first to observe that Construction 1.2 improves the efficiency of signing using pre-computation, as discussed at the end of Section 1.7.2. Essentially the same construction used to prove Theorem 1.1 was proposed and proven secure by Cramer and Pedersen [35, 32], and later re-investigated by [86] in the context of some specific signature schemes. (Even, et al. [45] also show a construction of a CMA-secure signature scheme from an RMA-secure scheme, but their construction is less efficient than the one given in this chapter.) The construction in Section 1.8 is due to Bellare and Shoup [12]. Other proofs of a similar result appear in [68, 106].

Chapter 2
Cryptographic Hardness Assumptions

As noted in the previous chapter, it is impossible to construct a digital signature scheme that is secure against an all-powerful adversary. Instead, the best we can hope for is to construct schemes that are secure against *computationally bounded* adversaries (that, for our purposes, means adversaries running in probabilistic polynomial time). Even for this "limited" class of adversaries, however, we do not currently have any constructions that can proven, unconditionally, to be secure. In fact, it is not too difficult to see that the existence of a secure signature scheme would imply[1] $P \neq NP$, a breakthrough in complexity theory. (While there is general belief that $P \neq NP$ is true, we seem very far away from being able to prove this.) Actually, as we will see below, the existence of a secure signature scheme implies the existence of one-way functions, something not known to follow from $P \neq NP$ and thus an even stronger result. (Informally, the issue is that $P \neq NP$ only guarantees the existence of problems that are hard *in the worst case*. But a secure signature scheme is required to be "hard to break" *on the average* — in particular, for "average" public keys generated by signers.)

Given this state of affairs, all existing constructions of signature schemes are proven secure relative to some *assumption* regarding the hardness of solving some (cryptographic) problem. We introduce some longstanding and widely used assumptions in this chapter; other, more recent cryptographic assumptions are introduced as needed throughout the rest of the book.

2.1 "Generic" Cryptographic Assumptions

We begin by discussing "generic" cryptographic assumptions before turning to various concrete, number-theoretic constructions conjectured to satisfy these assump-

---

[1] See any book on complexity theory for definitions of these classes, which are not essential to the rest of the book.

J. Katz, *Digital Signatures*, DOI 10.1007/978-0-387-27712-7_2,
© Springer Science+Business Media, LLC 2010

tions. While any scheme used in practice must be based on some concrete "hard" problem, there are several advantages of studying generic assumptions:

- A signature scheme based on generic assumptions is not tied to any *particular* "hard" problem, and therefore offers greater flexibility to the implementor. As but one illustration of this flexibility, note that a signature scheme based on a *specific* assumption must be scrapped if the assumption is found to be false, whereas a signature scheme based on generic assumptions (that was instantiated with the particular assumption found to be false) can simply be instantiated using a different concrete problem.
- Constructions based on generic assumptions are often simpler to analyze and understand, since abstracting away the "unnecessary" details has the effect of highlighting those details that are important.
- Generic constructions are interesting from a theoretical point of view insofar as they indicate what is feasible, and what are the minimal assumptions that are necessary. These are often useful steps toward developing practical schemes.

On the other hand, tailoring a signature scheme to a specific assumption can often lead to a much more efficient scheme than simply "plugging in" that same assumption to a generic template. Indeed, constructions based on specific assumptions are generally orders of magnitude more efficient than schemes based on generic assumptions, regardless of how the latter are instantiated.

2.1.1 One-Way Functions and Permutations

The most basic building block in cryptography — in fact, as we will see, the minimal assumption needed for constructing secure signature schemes — is a *one-way function*. Informally, a one-way function f is a function that is "easy" to compute but "hard" to invert *on the average*, in a way made precise below. We give two definitions of a one-way function: the first is easier to work with, while the second is easier to instantiate using known number-theoretic primitives (and can also yield more efficient constructions). Fortunately, the two definitions are equivalent in the sense that one exists if and only if the other does. We also define two notions of one-way *permutations*, though equivalence in this case is not known to hold.

Definition 2.1. A function $f : \{0,1\}^* \to \{0,1\}^*$ is a **one-way function** if:

1. (Easy to compute:) There is a deterministic, polynomial-time algorithm Eval_f such that for all k and all $x \in \{0,1\}^k$ we have $\mathrm{Eval}_f(x) = f(x)$. (From now on, we do not explicitly mention Eval_f and only refer to f itself, keeping in mind that f can be computed in polynomial time.)
2. (Hard to invert:) The following is negligible for all PPT algorithms A:

$$\Pr\left[x \leftarrow \{0,1\}^k; y := f(x); x' \leftarrow A(1^k, y) : f(x') = y\right].$$

Note that it is not required that $x = x'$ in the above.

A one-way function f is a **one-way permutation** if f is bijective and length-preserving (i.e., $|f(x)| = |x|$ for all x).

It is not difficult to show that the existence of a one-way function implies $P \neq NP$. On the other hand, we do not currently know whether $P \neq NP$ necessarily implies the existence of one-way functions.

While the above definition is convenient when one-way functions or permutations are used to construct other objects, it does not provide a natural model for the concrete examples of one-way functions/permutations that we currently know. Instead, it is often simpler to work with *families* of one-way functions/permutations, defined next.

Definition 2.2. A tuple $\Pi = (\mathsf{Gen}, \mathsf{Samp}, f)$ of PPT algorithms is a **function family** if the following hold:

1. Gen, the **parameter-generation algorithm**, is a probabilistic algorithm that takes as input a security parameter 1^k and outputs parameters I with $|I| \geq k$. Each value of I output by Gen defines sets D_I and R_I that constitute the domain and range, respectively, of the function f_I defined below.
2. Samp, the **sampling algorithm**, is a probabilistic algorithm that takes as input parameters I and outputs an element of D_I (except possibly with negligible probability).
3. f, the **evaluation algorithm**, is a deterministic algorithm that takes as input parameters I and an element $x \in D_I$, and outputs an element $y \in R_I$. We write this as $y := f_I(x)$. (That is, the function f_I is defined by the behavior of f on parameters I.)

Π is a **permutation family** if the following additionally hold:

1. For all I output by Gen, the distribution defined by the output of $\mathsf{Samp}(I)$ is (statistically close to) the uniform distribution on D_I.
2. For all I output by Gen it holds that $D_I = R_I$ and the function f_I is a bijection.

If Π is a permutation family and there exists a polynomial p such that $D_I = \{0,1\}^{p(k)}$ for all I output by $\mathsf{Gen}(1^k)$, then we say that Π is a **permutation family over bit-strings**. In this case we will assume the trivial sampling algorithm (that simply outputs its random coins).

Definition 2.3. A function/permutation family $\Pi = (\mathsf{Gen}, \mathsf{Samp}, \mathsf{Eval})$ is **one-way** if for all PPT algorithms A, the following is negligible (in k):

$$\Pr[I \leftarrow \mathsf{Gen}(1^k); x \leftarrow \mathsf{Samp}(I); y := f_I(x); x' \leftarrow A(I, y) : f_I(x') = y].$$

Any one-way permutation family satisfying some mild additional conditions can be transformed into a one-way permutation family over bit-strings, and we now sketch this transformation. Let Π be a one-way permutation family with $D_I \subseteq \{0,1\}^{p(k)}$ (for some polynomial p) for all I output by $\mathsf{Gen}(1^k)$. We additionally require that:

- Given I, the set D_I is efficiently recognizable. (I.e., there is a polynomial-time algorithm A that takes as input I and a string $x \in \{0,1\}^{p(k)}$ and correctly decides whether $I \in D_I$.)
- For all I, the set D_I is *dense* in $\{0,1\}^{p(k)}$. That is, $|D_I|/2^{p(k)} = 1/\text{poly}(k)$.

Construct a permutation family $\Pi' = (\text{Gen}', \text{Samp}', f')$ as follows: Gen' is identical to Gen. The sampling algorithm Samp' is the trivial one that outputs a random string of length $p(k)$ (we assume that k can be determined from I). Finally, define function $f_I' : \{0,1\}^{p(k)} \to \{0,1\}^{p(k)}$ as:

$$f_I'(x) = \begin{cases} f_I(x) & x \in D_I \\ x & \text{otherwise} \end{cases}.$$

Note that Π' is not necessarily one-way, since f_I' is trivial to invert on any point $y \notin D_I$. Nevertheless, it *is* hard to invert f_I' on a noticeable fraction of its range. This hardness can be "amplified" by running many copes of Π' in parallel. That is, define $\Pi'' = (\text{Gen}'', \text{Samp}'', f'')$ as follows: Gen'' is the same as Gen. The sampling algorithm Samp' outputs a random string of length $\ell(k) \cdot p(k)$ for an appropriate polynomial ℓ. Finally,

$$f_I''(x_1 \| \cdots \| x_{\ell(k)}) \stackrel{\text{def}}{=} f_I'(x_1) \| \cdots \| f_I'(x_{\ell(k)}).$$

Intuitively, it is clear that inversion is difficult as long as *any* of the x_i are in D_I, and this is true for some x_i with all but negligible probability (for ℓ chosen appropriately). A formal proof that Π'' is a one-way permutation family over bit-strings is not much more difficult.

We have defined both one-way functions (cf. Definition 2.1) and one-way function families (cf. Definition 2.3). We now show that these definitions are equivalent.

Lemma 2.1. *A one-way function f (in the sense of Definition 2.1) exists iff a one-way function family (in the sense of Definition 2.3) exists.*

Proof (sketch). It is immediate that a one-way function f implies the existence of a one-way function family: simply let Gen be the trivial algorithm that on input 1^k outputs $I = 1^k$; take Samp to be the algorithm that on input $I = 1^k$ outputs a uniformly distributed string $x \in \{0,1\}^k$; and define $f_I(x) = f(x)$.

The other direction is also conceptually simple, just more technical. Let $\Pi = (\text{Gen}, \text{Samp}, f)$ be a one-way function family such that the running time of Gen is bounded by p_1 and the running time of Samp is bounded by p_2, and let $p \stackrel{\text{def}}{=} p_1 + p_2$; note that p is a polynomial and furthermore that the combined length of the random tapes used by Gen and Samp for security parameter k is bounded by $p(k)$. Define f as follows: on input $r \in \{0,1\}^k$ find the largest integer $\bar{k}$ such that $p(\bar{k}) \leq k$. Parse r as $r_1 | r_2$ with $|r_1| = p_1(\bar{k})$ and $|r_2| \geq p_2(\bar{k})$. Set $I := \text{Gen}(1^{\bar{k}}; r_1)$ and $x := \text{Samp}(I; r_2)$ (note that we fix the random tapes of Gen and Samp, so this step is deterministic), and compute $y := f_I(x)$. The output of f is the pair (I, y). The proof that f is a one-way function is tedious, but straightforward.

The above shows that one-way functions are equivalent to one-way function families. In contrast, while the existence of one-way permutations is easily seen to imply the existence of one-way permutation families, the converse is not known. Moreover, the specific number-theoretic assumptions discussed below yield one-way permutation families (indeed, one-way permutation families over bit-strings) much more naturally than they do one-way permutations. We will therefore work exclusively with the notion of one-way permutation families over bit-strings.

This is a good place to record the following observation.

Theorem 2.1. *The existence of a signature scheme that is existentially unforgeable under a one-time random-message attack implies the existence of a one-way function.*

Proof (sketch). In fact even security against a *no*-message attack suffices to prove the claim. Let $\Pi = (\mathsf{Gen}, \mathsf{Sign}, \mathsf{Vrfy})$ be a signature scheme that is existentially unforgeable under a no-message attack, where an adversary is given only the public key pk and succeeds if it outputs (m, σ) with $\mathsf{Vrfy}_{pk}(m, \sigma) = 1$. Let $p(k)$ be a polynomial bounding the length of the random tape used by Gen on security parameter 1^k. Define a one-way function f as follows: on input $r \in \{0, 1\}^k$, compute the largest integer $\bar{k}$ such that $p(\bar{k}) \leq k$. Then run $\mathsf{Gen}(1^{\bar{k}}; r)$ to obtain (pk, sk), and output pk.

Observe that any PPT algorithm A inverting f can be used to forge signatures in Π as follows: given pk, run A to obtain a string r. If $f(r) = pk$, then this means that running $\mathsf{Gen}(1^{\bar{k}}; r)$ yields a pair (pk, sk'). It is then trivial to output a forgery on any message m by computing the signature $\sigma \leftarrow \mathsf{Sign}_{sk'}(m)$. (Note that sk' need not be equal to the "real" secret key sk used by the signer; i.e., there may be multiple valid secret keys associated with the single public key pk. But correctness of Π implies that this does not matter, since valid signatures with respect to pk can be produced using *any* secret key associated with pk.)

2.1.2 Trapdoor Permutations

A stronger notion than that of one-way functions is obtained by introducing an "asymmetry" of sorts whereby one party *can* efficiently accomplish some task that is infeasible for anyone else. This leads to the idea of *trapdoor permutations* that may be viewed, informally, as one-way permutations that can be efficiently inverted given some additional "trapdoor" information. (One can also consider trapdoor *functions* but these turn out to be much less useful.) A definition follows.

Definition 2.4. A tuple $\Pi = (\mathsf{Gen}, \mathsf{Samp}, f, f^{-1})$ of PPT algorithms is a **trapdoor permutation family** if the following hold:

- Gen, the **parameter-generation algorithm**, is a probabilistic algorithm that takes as input a security parameter 1^k and outputs parameters I (with $|I| \geq k$) along with an associated trapdoor td.

Each value of I output by Gen defines a set D_I that constitutes the domain and range of a permutation f_I defined below.

- Samp, the **sampling algorithm**, is a probabilistic algorithm that takes as input parameters I and outputs an element $x \in D_I$ whose distribution is statistically close to the uniform distribution over D_I. We will sometimes leave Samp implicit and just write[2] $x \leftarrow D_I$.

- f, the **evaluation algorithm**, is a deterministic algorithm that takes as input parameters I and an element $x \in D_I$, and outputs an element $y \in D_I$. We write this as $y := f_I(x)$.

- f^{-1}, the **inversion algorithm**, is a deterministic algorithm that takes as input parameters I, a trapdoor td, and an element $y \in D_I$. It outputs an element $x \in D_I$. We leave I implicit, and write this as $x := f_{\mathsf{td}}^{-1}(y)$.

- For all k, all (I, td) output by $\mathsf{Gen}(1^k)$, and all $x \in D_I$ we have $f_{\mathsf{td}}^{-1}(f_I(x)) = x$, and hence $f_{\mathsf{td}}^{-1}(\cdot)$ and $f_I(\cdot)$ are both permutations over D_I, and inverses of each other.

- The following is negligible for all PPT algorithms A:

$$\Pr\left[(I, \mathsf{td}) \leftarrow \mathsf{Gen}(1^k); y \leftarrow D_I; x \leftarrow A(I, y) : f_I(x) = y\right].$$

For brevity, and since it will not cause confusion, we simply refer to a "trapdoor permutation" rather than a "trapdoor permutation family".

Because f_I is a permutation, choosing x uniformly from D_I and then setting $y := f_I(x)$ results in a value y that is uniformly distributed in D_I. We note also that it is possible for f_I to be defined over some set that (strictly) contains D_I, but the function is only guaranteed to be a *bijection* when its inputs are taken from D_I. The final condition of the definition, however, requires that it be "hard" to find *any* x mapping to y (i.e., even an $x \notin D_I$).

Occasionally, when we do not care about the particular index I or trapdoor td, we will write $(f, f^{-1}) \leftarrow \mathsf{Gen}(1^k)$ and write $f(\cdot)$ in place of $f_I(\cdot)$ and $f^{-1}(\cdot)$ in place of f_{td}^{-1}. Of course, it is important to keep in mind that I is required in order to evaluate f, and that f^{-1} can only be evaluated efficiently with knowledge of td.

Trapdoor permutations, in the sense defined above, do not suffice for most of the applications we will see in this book. Instead, we need the following strengthening:

Definition 2.5. A trapdoor permutation family $\Pi = (\mathsf{Gen}, \mathsf{Samp}, f, f^{-1})$ is called **doubly enhanced**[3] if the following conditions hold:

1. The following is negligible for all PPT algorithms A:

$$\Pr\left[(I, \mathsf{td}) \leftarrow \mathsf{Gen}(1^k); r \leftarrow \{0, 1\}^*; y := \mathsf{Samp}(I; r); x \leftarrow A(I, y, r) : f_I(x) = y\right].$$

---

[2] Technically, $x \leftarrow D_I$ refers to selecting x uniformly from D_I. Since the distribution produced by Samp is statistically close to uniform, the difference is unimportant.

[3] We use this terminology to distinguish our definition from that of enhanced trapdoor permutations, which satisfy only the first condition.

That is, it should be difficult to find a pre-image of y *even when given the random coins used to sample y*.

2. Let $p(k)$ denote the length of the random tape used by Samp on security parameter 1^k. There exists a PPT algorithm Samp$'$ that takes as input I and outputs a tuple (x, y, r) with $x \in D_I$ and $r \in \{0, 1\}^{p(k)}$ and such that:

   - $f_I(x) = y$ and $y = \mathsf{Samp}(I; r)$;
   - The distribution on r is statistically close to uniform.

We can also define a trapdoor permutation *over bit-strings* in the natural way (cf. Definition 2.2). It is not hard to see that any trapdoor permutation over bit-strings is also a doubly enhanced trapdoor permutation: the first condition of Definition 2.5 holds by virtue of the fact that Samp is trivial (since $y = \mathsf{Samp}(I; y)$), and the second condition holds by letting Samp$'$ be the algorithm that chooses x uniformly, sets $y := f_I(x)$, and sets $r := y$. All the concrete examples of trapdoor permutations that we will see in this book can be suitably "massaged" to be trapdoor permutations over bit-strings.

2.1.3 Clawfree (Trapdoor) Permutations

A pair of *clawfree permutations* is, informally, a pair of efficiently computable permutations f_0, f_1 defined over the same domain for which it is hard to find a *claw*: namely, a pair x_0, x_1 with $f_0(x_0) = f_1(x_1)$. A pair of clawfree *trapdoor* permutations additionally has an associated *trapdoor* td that allows for efficient inversion of f_0 and f_1. Observe that given such trapdoor information, it is easy to find a claw: simply choose arbitrary y and compute $x_0 := f_0^{-1}(y)$ and $x_1 := f_1^{-1}(y)$; thus, hardness of finding a claw holds only for algorithms that do not have access to the trapdoor.

Definition 2.6. A tuple $\Pi = (\mathsf{Gen}, \mathsf{Samp}, f_0, f_1)$ of PPT algorithms is a **clawfree permutation family** if the following hold:

- Gen, the **parameter-generation algorithm**, is a probabilistic algorithm that takes as input a security parameter 1^k and outputs parameters I (with $|I| \geq k$) along with an associated trapdoor td.
  Each value of I output by Gen defines a set D_I that constitutes the domain and range of permutations $f_{I,0}, f_{I,1}$ defined below.
- Samp, the **sampling algorithm**, is a probabilistic algorithm that takes as input parameters I and outputs an element $x \in D_I$ whose distribution is statistically close to the uniform distribution over D_I. We usually leave Samp implicit, and just write $x \leftarrow D_I$.
- f_0 and f_1, the **evaluation algorithms**, are deterministic algorithms that take as input parameters I and an element $x \in D_I$, and output an element $y \in D_I$. We write this as $y := f_{I,0}(x)$ or $y := f_{I,1}(x)$.

- The following is negligible for all PPT algorithms A:

$$\Pr\left[(I,\mathsf{td}) \leftarrow \mathsf{Gen}(1^k); (x_0,x_1) \leftarrow A(I) : f_{I,0}(x_0) = f_{I,1}(x_1)\right].$$

$\Pi = (\mathsf{Gen},\mathsf{Samp},f_0,f_1,f_0^{-1},f_1^{-1})$ is a **clawfree trapdoor permutation family** if $(\mathsf{Gen},\mathsf{Samp},f_0,f_1)$ is a clawfree permutation family and the following additionally hold:

- f_0^{-1} and f_1^{-1}, the **inversion algorithms**, are deterministic algorithms that take as input parameters I, a trapdoor td, and an element $y \in D_I$. They output an element $x \in D_I$. We leave I implicit, and write this as $x := f_{\mathsf{td},0}^{-1}(y)$ or $x := f_{\mathsf{td},1}^{-1}(y)$.
- For all k, all (I,td) output by $\mathsf{Gen}(1^k)$, all $x \in D_I$, and $b \in \{0,1\}$ we have $f_{\mathsf{td},b}^{-1}(f_{I,b}(x)) = x$. Thus, $f_{\mathsf{td},b}^{-1}(\cdot)$ and $f_{I,b}(\cdot)$ are permutations over D_I and inverses of each other.

As in the case of trapdoor permutations, we often refer to "clawfree (trapdoor) permutations" rather than "clawfree (trapdoor) permutation families." We may also switch to a less cumbersome notation and write $(f_0,f_1,f_0^{-1},f_1^{-1}) \leftarrow \mathsf{Gen}(1^k)$ for the output of Gen, and then use $f_0(\cdot),f_1(\cdot)$ in place of $f_{I,0}(\cdot),f_{I,1}(\cdot)$ and, similarly, use $f_0^{-1}(\cdot),f_1^{-1}(\cdot)$ in place of $f_{\mathsf{td},0}^{-1}(\cdot),f_{\mathsf{td},1}^{-1}(\cdot)$. As before, it is important to keep in mind that f_0^{-1},f_1^{-1} cannot be efficiently evaluated without knowledge of td.

We also note, once again, that it is possible for f_0,f_1 to be defined over some set (strictly) containing the domain D over which these functions are guaranteed to be permutations. The final condition of the definition, however, requires that it be "hard" to find *any* x_0,x_1 for which $f_0(x_0) = f_1(x_1)$ (i.e., even $x_0,x_1 \notin D$).

The existence of clawfree trapdoor permutations represents a (possibly) stronger assumption than the existence of trapdoor permutations:

Lemma 2.2. *If* $\Pi = (\mathsf{Gen},\mathsf{Samp},f_0,f_1,f_0^{-1},f_1^{-1})$ *is a clawfree trapdoor permutation family, then* $\Pi' = (\mathsf{Gen},\mathsf{Samp},f_0,f_0^{-1})$ *is a trapdoor permutation family. Thus, the existence of clawfree permutations implies the existence of trapdoor permutations.*

Proof (sketch). The syntactic requirements are easily seen to match up, and so all we need to prove is hardness of inversion. Fix any PPT algorithm A' and define:

$$\delta_{A'}(k) \overset{\text{def}}{=} \Pr\left[(I,\mathsf{td}) \leftarrow \mathsf{Gen}(1^k); y \leftarrow D_I; x \leftarrow A'(I,y) : f_0(x) = y\right].$$

This is exactly the probability with which A' inverts Π', and so we prove that Π' is a trapdoor permutation family by showing that $\delta_{A'}(k)$ is negligible.

Consider the following algorithm A for finding a claw in Π, using A' as a subroutine:

> **Algorithm A:**
> The algorithm is given parameters I, generated using $\mathsf{Gen}(1^k)$.
> Its goal is to find a claw.

- Choose $x_1 \leftarrow D_I$ and compute $y := f_1(x_1)$.
- Run $A'(I, y)$ to obtain x_0.
- If $f_0(x_0) = y$, then output the claw (x_0, x_1).

Clearly, A runs in polynomial time. Furthermore, A succeeds in outputting a claw whenever A' succeeds in inverting y with respect to f_0. Since A chooses x_1 uniformly at random from D_I and f_1 is a permutation over this set, the value y given by A to A' is also uniformly distributed in D_I. Thus, the probability that A' succeeds in inverting y is exactly $\delta_{A'}(k)$, and this is exactly the probability with which A outputs a claw. The fact that Π is clawfree thus implies that $\delta_{A'}(k)$ is negligible, as desired.

By analogy with the case of trapdoor permutations, we may also define a notion of doubly enhanced clawfree trapdoor permutations:

Definition 2.7. Let $\Pi = (\mathsf{Gen}, \mathsf{Samp}, f_0, f_1, f_0^{-1}, f_1^{-1})$ be a clawfree trapdoor permutation family. We say Π is **doubly enhanced** if both $\Pi_0 = (\mathsf{Gen}, \mathsf{Samp}, f_0, f_0^{-1})$ and $\Pi_1 = (\mathsf{Gen}, \mathsf{Samp}, f_1, f_1^{-1})$ are doubly enhanced trapdoor permutation families. That is:

- For $b \in \{0, 1\}$ and any PPT algorithm A the following is negligible:

$$\Pr\left[(I, \mathsf{td}) \leftarrow \mathsf{Gen}(1^k); r \leftarrow \{0, 1\}^*; y := \mathsf{Samp}(I; r); x \leftarrow A(I, y, r) : f_{I,b}(x) = y\right].$$

- If we let $p(k)$ denote the length of the random tape used by Samp on security parameter 1^k, there exist PPT algorithms $\mathsf{Samp}_0, \mathsf{Samp}_1$ where Samp_b takes as input I and outputs a tuple (x, y, r) with $x \in D_I$ and $r \in \{0, 1\}^{p(k)}$ and such that:

1. $f_{I,b}(x) = y$ and $y = \mathsf{Samp}(I; r)$;
2. the distribution on r is statistically close to uniform.

We may also define a *clawfree trapdoor permutation family over bit-strings* in the obvious way, and it is easy see that any such family is also a doubly enhanced clawfree trapdoor permutation family.

2.2 Specific Assumptions

The discussion thus far has been very general. We now show some concrete examples of number-theoretic problems conjectured to be hard, and demonstrate how these can be used to instantiate the generic assumptions described thus far. We assume in this section some familiarity with basic number theory; see the notes at the end of this chapter for pointers to existing references covering this material.

In this chapter we have chosen to focus on the most well known and long-standing cryptographic assumptions; some more recent assumptions are introduced and discussed in Chapters 4 and 5.

2.2.1 Hardness of Factoring

The factoring problem is probably the longest-studied "hard" problem in algorithmic number theory. It is also one of the easiest one-way functions to explain, at least informally, since multiplication is clearly "easy" (i.e., polynomial time) yet finding the prime factorization of a (large) number is widely believed[4] to be "hard". But does the conjectured "hardness of factoring" trivially imply a one-way function? A natural first candidate for a one-way function is the function $f_{mult}(x,y) = xy$. A little thought, however, shows that f_{mult} is decidedly *not* one-way: with probability $3/4$ at least one of x or y will be even, making it trivial to find a factor of xy (recall that one-wayness is defined in terms of the inability to find *any* preimage of a randomly generated point). To avoid problems of this sort, we simply need to restrict the inputs of f_{mult} to (large) *primes* of equal length. Formally, we construct a function family (Gen, Samp, f) as follows (cf. Definition 2.3):

- Gen(1^k) simply outputs $I = 1^k$. We let D_I denote the set of all pairs of k-bit primes.
- Samp(1^k) is a randomized algorithm that outputs two random (and independently chosen) k-bit primes.
- $f(p,q)$ outputs the product pq.

One way to state the factoring conjecture is as the assumption that the family (Gen, Samp, f) defined above is one-way.

Of course, we have omitted what is perhaps the most important detail in the above: how to generate random primes in polynomial time. An algorithm computing Samp follows fairly readily from the following two facts:

1. Prime numbers are sufficiently dense that a random integer is prime with "sufficiently high" probability.
2. There exist (probabilistic) polynomial-time algorithms that can determine (except with negligibly small error) whether a given integer is prime.

We refer the reader to the references listed in the notes at the end of this chapter for further information.

For our purposes, it will be convenient to let GenModulus denote an (unspecified, but polynomial-time) algorithm that, on input 1^k, outputs (N, p, q) such that $N = pq$, and p and q are k-bit primes (with all but negligible probability in k). We can then express the factoring assumption relative to a particular algorithm GenModulus:

Definition 2.8. We say that **factoring is hard relative to** GenModulus if for all PPT algorithms A, the following is negligible:

$$\Pr[(N,p,q) \leftarrow \mathsf{GenModulus}(1^k); (p,q) \leftarrow A(N) : pq = N].$$

---

[4] It is crucial to keep in mind here that running time is measured in terms of the *length(s) of the input(s)* and not their magnitude. It is easy to factor a number N in time linear in N using trial division by all numbers less than N. But a polynomial-time algorithm for factoring N is required to work in time polynomial in $|N| = \Theta(\log N)$.

The factoring assumption is that there exists a GenModulus relative to which factoring is hard.

We stress that we do not require that GenModulus choose p and q to be *random* k-bit primes; though that is certainly one possibility (that is also used frequently in practice), we allow for other means of choosing the primes p and q so long as the factoring assumption (relative to GenModulus) is still believed to hold.

Interestingly, the factoring assumption — that, at first glance, seems only to guarantee the existence of a one-way function — can be used to construct a much stronger cryptographic primitive: a (doubly enhanced) clawfree trapdoor permutation family. We first show how to use the factoring assumption to construct a trapdoor permutation family, and then describe the extension to give the result claimed.

We begin with a small amount of (standard) number-theoretic background. Given any integer $N > 1$, let $\mathbb{Z}_N \overset{\text{def}}{=} \{0, \ldots, N-1\}$. It is a well-known fact that this is a group under addition modulo N. We also define

$$\mathbb{Z}_N^* \overset{\text{def}}{=} \{x \in \{1, \ldots, N-1\} \mid \gcd(x, N) = 1\}.$$

It is not too difficult to prove that $\mathbb{Z}_N^*$ is a group under multiplication modulo N; this follows from the fact that $\mathbb{Z}_N^*$ contains exactly those elements of $\mathbb{Z}_N$ that have a multiplicative inverse modulo N.

The squaring function modulo N is the function that maps $x \in \mathbb{Z}_N^*$ to $x^2 \bmod N$. Elements of $\mathbb{Z}_N^*$ that have a square root are called *quadratic residues* modulo N, and we denote the set of quadratic residues modulo N by QR_N. If N is a product of two distinct, odd primes, then squaring modulo N is a four-to-one function; i.e., each quadratic residue modulo N has exactly four square roots. We use this fact in the proof of the following:

Lemma 2.3. *Let $N = pq$ with p, q distinct, odd primes. Given $x, \hat{x}$ such that $x^2 = y = \hat{x}^2 \bmod N$ but $x \neq \pm\hat{x} \bmod N$, it is possible to factor N in polynomial time.*

Proof. We claim that either $\gcd(N, x + \hat{x})$ or $\gcd(N, x - \hat{x})$ is equal to one of the prime factors of N. Since gcd computations can be carried out in polynomial time, this proves the lemma.

If $x^2 = \hat{x}^2 \bmod N$ then

$$0 = x^2 - \hat{x}^2 = (x - \hat{x}) \cdot (x + \hat{x}) \bmod N,$$

and so $N \mid (x - \hat{x})(x + \hat{x})$. Then $p \mid (x - \hat{x})(x + \hat{x})$ and so p divides one of these terms. Say $p \mid (x + \hat{x})$ (the proof proceeds similarly if $p \mid (x - \hat{x})$). If $q \mid (x + \hat{x})$ then $N \mid (x + \hat{x})$, but this cannot be the case since $x \neq -\hat{x} \bmod N$. So $q \nmid x + \hat{x}$ and $\gcd(N, x + \hat{x}) = p$.

The following important result shows (roughly) that if N is hard to factor then squaring modulo N is one-way. Formally, define a function family $\Pi_{\mathsf{squaring}} = (\mathsf{Gen}, \mathsf{Samp}, f)$ as follows:

- $\mathsf{Gen}(1^k)$ computes $(N, p, q) \leftarrow \mathsf{GenModulus}(1^k)$, and outputs parameters N. Let $D_N = \mathbb{Z}_N^*$ and $R_N = \mathsf{QR}_N$.

- Samp(N) chooses a uniform element of $\mathbb{Z}_N^*$. (This can be done easily by choosing random elements of $\mathbb{Z}_N$ until one is found that is relatively prime to N.)
- $f_N(x)$ outputs $x^2 \bmod N$.

Theorem 2.2. *If factoring is hard relative to* GenModulus, *then* Π_{squaring} *is a one-way function family.*

Proof. Let A be a probabilistic polynomial-time algorithm, and define

$$\varepsilon_A(k) \stackrel{\text{def}}{=} \Pr\left[N \leftarrow \text{Gen}(1^k); y \leftarrow \text{QR}_N; x \leftarrow A(N,y) : x^2 = y \bmod N\right].$$

Since setting $y := x^2 \bmod N$ for a uniformly random $x \in \mathbb{Z}_N^*$ is equivalent to choosing y uniformly from QR_N (because squaring is four-to-one), the above exactly represents A's success probability in inverting the squaring function modulo N. Showing that $\varepsilon_A(k)$ is negligible thus proves the theorem.

Consider the following probabilistic polynomial-time algorithm A_{fact} that attempts to factor moduli output by GenModulus:

> **Algorithm A_{fact}:**
> The algorithm is given a modulus N as input.
>
> - Choose random $x \leftarrow \mathbb{Z}_N^*$ and compute $y := x^2 \bmod N$.
> - Run $A(N,y)$ to obtain output $\hat{x}$.
> - If $\hat{x}^2 = y \bmod N$ and $\hat{x} \neq \pm x \bmod N$, then factor N using Lemma 2.3.

By Lemma 2.3, we know that A_{fact} succeeds in factoring N exactly when $\hat{x} \neq \pm x \bmod N$ and $\hat{x}^2 = y \bmod N$. Since the modulus N given as input to A_{fact} is generated by GenModulus(1^k), and y is a random quadratic residue modulo N (since x was chosen uniformly at random from $\mathbb{Z}_N^*$), the probability that A outputs $\hat{x}$ satisfying $\hat{x}^2 = y \bmod N$ is exactly $\varepsilon_A(k)$. Moreover, conditioned on the value of the quadratic residue y given to A, the value x used by A_{fact} is equally likely to be any of the four possible square roots of y. This means that, conditioned on A outputting *some* square root $\hat{x}$ of y, the probability that $\hat{x} \neq \pm x \bmod N$ is exactly $1/2$. Putting this together, we have:

$$\Pr[(N,p,q) \leftarrow \text{GenModulus}(1^k) : A_{\text{fact}} \text{ factors } N]$$

$$= \Pr\left[\begin{array}{c} (N,p,q) \leftarrow \text{GenModulus}(1^k); x \leftarrow \mathbb{Z}_N^*; \\ y := x^2 \bmod N; \hat{x} \leftarrow A(N,y) \end{array} : \hat{x} \neq \pm x \bmod N \bigwedge \hat{x}^2 = y \bmod N\right]$$

$$= \frac{1}{2} \cdot \Pr\left[\begin{array}{c} (N,p,q) \leftarrow \text{GenModulus}(1^k); x \leftarrow \mathbb{Z}_N^*; \\ y := x^2 \bmod N; \hat{x} \leftarrow A(N,y) \end{array} : \hat{x}^2 = y \bmod N\right]$$

$$= \varepsilon_A(k)/2.$$

Since factoring is hard relative to GenModulus, we conclude that $\varepsilon_A(k)$ must be negligible, completing the proof.

One approach to making Π_{squaring} a permutation family is to consider specific moduli N and restrict the domain of the function. For $N = pq$ a product of two distinct primes p and q, we say that N is a *Blum integer* if $p = q = 3 \bmod 4$. It is a fact that if N is a Blum integer, then any quadratic residue modulo N has exactly one square root that is *also* a quadratic residue. Thus, the squaring function for a Blum integer N is a permutation over QR_N.

It is also known that computing square roots modulo N is "easy" (i.e., can be done in polynomial time) given the factorization of N. Combining this with the previous observation, we obtain a trapdoor permutation family based on factoring:

- $\text{Gen}(1^k)$ computes $(N, p, q) \leftarrow \text{GenModulus}(1^k)$, where GenModulus is such that $p = q = 3 \bmod 4$. It then outputs parameters N and trapdoor (p, q). Let $D_N = \text{QR}_N$.
- $\text{Samp}(N)$ chooses a uniform element of $y \in \text{QR}_N$. (This can be done easily by choosing a random element $r \in \mathbb{Z}_N^*$ and setting $y := r^2 \bmod N$.)
- $f_N(x)$ outputs $x^2 \bmod N$.
- $f_{(p,q)}^{-1}(y)$ computes the unique square root of y modulo N that is itself a quadratic residue.

Theorem 2.2 implies that the above is a trapdoor permutation family as long as factoring is hard relative to GenModulus. Notice, however, that (as described) it is not a *doubly enhanced* trapdoor permutation family: given the random coins r used by Samp, which we view[5] as an element of $\mathbb{Z}_N^*$, it is trivial to compute a square root of the output value $y = r^2 \bmod N$. We will see below how this can be addressed.

Extending the above gives a construction of a *clawfree* trapdoor permutation family:

- $\text{Gen}(1^k)$ computes $(N, p, q) \leftarrow \text{GenModulus}(1^k)$, where $p = q = 3 \bmod 4$, and chooses random $z \leftarrow \text{QR}_N$. It then outputs parameters (N, z) and trapdoor (p, q). Let $D_N = \text{QR}_N$.
- $\text{Samp}(N)$ chooses a uniform element of QR_N as above.
- Given (N, z), we define f_0 and f_1 as follows:

$$f_0(x) = x^2 \bmod N \quad \text{and} \quad f_1(x) = z \cdot x^2 \bmod N.$$

- Given (N, z) and the trapdoor information (p, q), the inverses of f_0 and f_1 can be computed as follows: To compute $f_0^{-1}(y)$, find the unique square root of y modulo N that is itself a quadratic residue. To compute $f_1^{-1}(y)$, find the unique square root of $y/z \bmod N$ that is itself a quadratic residue.

Theorem 2.3. *If factoring is hard relative to* GenModulus, *then the above constitutes a clawfree trapdoor permutation family.*

---

[5] This is justified since it is easy to map a sufficiently long bit-string to an element of $\mathbb{Z}_N^*$ such that a random bit-string yields an element of $\mathbb{Z}_N^*$ whose distribution is statistically close to uniform, and the mapping is invertible in the sense required.

Proof. The only condition difficult to verify is that it is computationally infeasible to find a claw. We show that any "claw-finding" algorithm A can be used to compute square roots modulo N. Theorem 2.2 thus implies that finding a claw is computationally infeasible.

Fix any PPT algorithm A and define

$$\varepsilon_A(k) \overset{\text{def}}{=} \Pr[(N,z,p,q) \leftarrow \mathsf{Gen}(1^k); (x_0,x_1) \leftarrow A(N,z) : x_0^2 = z \cdot x_1^2]. \qquad (2.1)$$

Since this is exactly the probability with which A succeeds in finding a claw, we need to show that $\varepsilon_A(k)$ is negligible.

Construct a PPT algorithm A' computing modular square roots as follows:

> **Algorithm A':**
> The algorithm is given a modulus N and an element $z \in \mathsf{QR}_N$
> as input.
>
> - Run $A(N,z)$ to obtain output (x_0,x_1).
> - If $x_0^2 = z \cdot x_1^2 \bmod N$, then output $x_0/x_1 \bmod N$.

It is easy to see that if the input z given to A' is chosen uniformly from QR_N, then the input (N,z) given to A is distributed identically to the experiment of Equation (2.1). Thus, the probability that A outputs (x_0,x_1) with $x_0^2 = z \cdot x_1^2 \bmod N$ is exactly $\varepsilon_A(k)$. Furthermore, whenever this occurs A' outputs a square root of its input z. But if factoring is hard relative to GenModulus, we know from Theorem 2.2 that this can happen with only negligible probability.

As in the case of the trapdoor permutation family presented earlier, the construction just given is not doubly enhanced. We will fix this below. To do so, we need to introduce some brief facts about the *Jacobi function* $\mathscr{J}_N : \mathbb{Z}_N^* \to \{-1,+1\}$. (We introduce here all the facts that are needed for the construction that follows. For further information about the Jacobi function, consult the references at the end of the chapter.) An element $x \in \mathbb{Z}_N^*$ with $\mathscr{J}_N(x) = +1$ is said to have *Jacobi symbol +1*, and similarly if $\mathscr{J}_N(x) = -1$ then we say that x has Jacobi symbol -1. The relevant facts are:

1. Exactly half the elements of $\mathbb{Z}_N^*$ have Jacobi symbol $+1$, and half have Jacobi symbol -1.
2. Given N,x, it is possible to compute $\mathscr{J}_N(x)$ in polynomial time *without knowledge of the factorization of N*.
3. For N a Blum integer, we have seen that every quadratic residue z has exactly one square root x that is also a quadratic residue. It is furthermore the case that z has exactly two square roots with Jacobi symbol $+1$, and these are given by $\pm x \bmod N$.

Let $\mathscr{J}_N^{+1}$ denote the set of elements of $\mathbb{Z}_N^*$ with Jacobi symbol $+1$. We now present the construction of a doubly enhanced clawfree trapdoor permutation:

- $\mathsf{Gen}(1^k)$ computes $(N, p, q) \leftarrow \mathsf{GenModulus}(1^k)$, where $p = q = 3 \bmod 4$, and chooses random $z \leftarrow \mathsf{QR}_N$. It then outputs parameters (N, z) and trapdoor (p, q). Let $D_N = \mathsf{QR}_N$.
- $\mathsf{Samp}(N)$ chooses a uniform element $y \in \mathsf{QR}_N$ by choosing a random $r \in \mathcal{J}_N^{+1}$ and setting $y := r^2 \bmod N$. (The random $r \in \mathcal{J}_N^{+1}$ is chosen by taking random bit-strings $r_1, \ldots$ of the appropriate length, and letting r be the first of these that is in $\mathbb{Z}_N^*$ and has Jacobi symbol $+1$.)
- Given (N, z), we define f_0 and f_1 as follows:

$$f_0(x) = x^4 \bmod N \quad \text{and} \quad f_1(x) = z^2 \cdot x^4 \bmod N.$$

- Given (N, z) and the trapdoor information (p, q), the inverses of f_0 and f_1 can be computed as follows: To compute $f_0^{-1}(y)$, find the unique fourth root of y modulo N that is itself a quadratic residue. To compute $f_1^{-1}(y)$, find the unique fourth root of $y/z^2 \bmod N$ that is itself a quadratic residue.

Theorem 2.4. *If factoring is hard relative to* $\mathsf{GenModulus}$, *then the above constitutes a doubly enhanced clawfree trapdoor permutation family.*

Proof (sketch). We first show that finding a claw implies the ability to compute square roots modulo N; it follows from Theorem 2.2 that finding a claw is infeasible. Say we are given N and a quadratic residue $z \in \mathsf{QR}_N$. Given a claw (x_0, x_1) such that $x_0^4 = z^2 \cdot x_1^4 \bmod N$, we have $(x_0^2/x_1^2)^2 = z^2 \bmod N$. Since the quadratic residue $z^2 \bmod N$ has a *unique* square root that is also a quadratic residue (namely, z itself), it must be the case that $x_0^2/x_1^2 = z \bmod N$ and so $x_0/x_1 \bmod N$ is a square root of z as desired.

A similar proof shows that it is hard to find a pre-image of y with respect to f_1 even when given the randomness used to generate y. Let A be a PPT algorithm inverting f_1, and consider the following probabilistic polynomial-time algorithm A' for computing square roots:

> **Algorithm A':**
> The algorithm is given a modulus N and $\hat{r} \in \mathsf{QR}_N$ as inputs.
>
> - Choose random $\hat{z} \in \mathbb{Z}_N^*$ and set $z := \hat{z}^2 \bmod N$.
> - Choose random $b \in \{0, 1\}$ and set $r := (-1)^b \hat{r} \bmod N$.
> - Compute $y := r^2 \bmod N$.
> - Run $A(N, z, y, r)$ to obtain output x.
> - Output $\hat{z} \cdot x$.

The input to A is distributed correctly, since r is uniformly distributed in $\mathcal{J}_N^{+1}$ (this follows because $\mathcal{J}_N(\hat{r}) = +1$ since $\hat{r}$ is a quadratic residue) and z is uniform in QR_N. Furthermore, we have

$$z^2 x^4 = y \bmod N \Rightarrow \left(\hat{z}^2 x^2\right)^2 = \hat{r}^2 \bmod N \Rightarrow \hat{z}^2 x^2 = \hat{r} \bmod N$$

(the final implication uses the fact that quadratic residues have a unique square root that is also a quadratic residue), and so $\hat{z}x$ is a square root of $\hat{r}$. Lemma 2.3 thus implies that A inverts with only negligible probability. A proof for the case of inverting f_0 follows analogously.

To conclude, we show algorithms $\mathsf{Samp}_0, \mathsf{Samp}_1$ as required[6] by Definition 2.7. Samp_0 proceeds as follows: Given (N, z), choose random $x \in \mathbb{Z}_N^*$ and compute $y := x^4 \bmod N$. Choose a random bit b and compute $r := (-1)^b \cdot x^2 \bmod N$. Output (x, y, r). It is clear that (x, y, r) satisfy the functional requirement of Definition 2.7. Furthermore, y is uniform in QR_N and r is a random square root of y having Jacobi symbol $+1$. From this it follows that r is uniform in $\mathcal{J}_N^{+1}$.

Algorithm Samp_1 as required by Definition 2.7 can be defined analogously as follows: Given (N, z), choose random $x \in \mathbb{Z}_N^*$ and compute $y := z^2 x^4 \bmod N$. Choose a random bit b and compute $r := (-1)^b \cdot z \cdot x^2 \bmod N$; output (x, y, r). Once again, (x, y, r) satisfy the functional requirement of Definition 2.7. Also, y is uniform in QR_N and r is a random square root of y having Jacobi symbol $+1$. From this it follows that r is uniform in $\mathcal{J}_N^{+1}$.

2.2.2 The RSA Assumption

Another popular assumption related to factoring is the *RSA assumption*, named after R. Rivest, A. Shamir, and L. Adleman who proposed this assumption in 1978. The RSA assumption implies that factoring is hard, while the converse it not known; thus, the RSA assumption is potentially stronger than the assumption that factoring is hard. (In other words, it is possible that an efficient algorithm for the RSA problem might be developed even while the factoring problem remains infeasible.) Nevertheless, the RSA assumption has stood the test of time for over 30 years.

We begin with a little background. Let $N = pq$ be a product of two odd primes p and q. Then we have seen that $\mathbb{Z}_N^*$ is a group with respect to multiplication modulo N. An easy computation shows that the number of elements in $\mathbb{Z}_N^*$ is given by $\phi(N) \stackrel{\text{def}}{=} (p-1) \cdot (q-1)$; in other words, $\phi(N)$ is the order of the group $\mathbb{Z}_N^*$. From this it follows that if e is an integer that is relatively prime to $\phi(N)$, then the function $f_{N,e} : \mathbb{Z}_N^* \to \mathbb{Z}_N^*$ given by $f_{N,e}(x) = x^e \bmod N$ is in fact a *permutation*. In other words, for every $y \in \mathbb{Z}_N^*$ there exists a unique $x \in \mathbb{Z}_N^*$ satisfying $x^e = y \bmod N$. For x, y satisfying this relation, we write $x = y^{1/e} \bmod N$.

The RSA problem, roughly speaking, is to compute the eth root of y modulo N; the RSA assumption is that computing eth roots modulo an integer N of unknown factorization is hard. Formally, let GenRSA be a probabilistic polynomial-time algorithm that, on input 1^k, outputs a modulus N that is the product of two k-bit primes (except possibly with negligible probability), along with an integer $e > 0$

---

[6] We concentrate on showing how a uniform $r \in \mathcal{J}_N^{+1}$ can be sampled, since going from this to the random coins needed to generate r is easy (due to the fact that the Jacobi symbol is efficiently computable).

with $\gcd(e, \phi(N)) = 1$ and an integer $d > 0$ satisfying $ed = 1 \bmod \phi(N)$. (We will see below the role played by d. Note that such a d is guaranteed to exist since e is relatively prime to $\phi(N)$.) Then:

Definition 2.9. We say that **the RSA problem is hard relative to** GenRSA if for all PPT algorithms A, the following is negligible:

$$\Pr[(N, e, d) \leftarrow \mathsf{GenRSA}(1^k); y \leftarrow \mathbb{Z}_N^*; x \leftarrow A(N, e, y) : x^e = y \bmod N].$$

The RSA assumption can be formalized as the assumption that there exists a GenRSA relative to which the RSA problem is hard. For certain applications, however, additional assumptions regarding the output of GenRSA are required.

The RSA assumption implies the existence of a one-way permutation family. Moreover, for any (N, e, d) output by GenRSA and any $y \in \mathbb{Z}_N^*$ we have

$$(y^d)^e = y^{de} = y^{de \bmod \phi(N)} = y^1 = y \bmod N,$$

and so computing eth roots is equivalent to raising to the dth power. Thus, viewing d as a trapdoor we see that the RSA assumption implies the existence of trapdoor permutations. Moreover, since sampling uniformly from $\mathbb{Z}_N^*$ is trivial, we actually obtain a doubly enhanced trapdoor permutation "for free".

As with GenModulus in the previous section, here too we are agnostic regarding the exact way GenRSA is implemented. One way to implement GenRSA based on any algorithm GenModulus (not necessarily outputting Blum integers) is as follows:

1. On input 1^k, compute $(N, p, q) \leftarrow \mathsf{GenModulus}(1^k)$. Then compute

$$\phi(N) := (p - 1)(q - 1).$$

2. Choose $e > 0$ such that $\gcd(e, \phi(N)) = 1$. (We leave the exact manner in which e is chosen unspecified, though in practice certain values of e may be preferred.) Compute $d := e^{-1} \bmod \phi(N)$ and output (N, e, d).

Since $\phi(N)$ can be computed given the factorization of N, and $d = e^{-1} \bmod \phi(N)$ can be computed in polynomial time given $\phi(N)$, it is clear that hardness of the RSA problem relative to GenRSA as constructed above implies hardness of factoring relative to GenModulus. As mentioned earlier, the converse is not known to be true. On the other hand, the only techniques currently known for attacking the RSA problem rely on first factoring N. In addition, it is known that recovering the trapdoor d, given N and e, *is* as hard as factoring (nevertheless, it is not clear that recovery of d is *necessary* in order to compute eth roots). With respect to existing technology, then, the RSA and factoring problems may be viewed as "equally hard".

As in the case of the factoring assumption, the RSA assumption may also be used to construct a (doubly enhanced) clawfree trapdoor permutation family in a very similar fashion:

- Gen(1^k) computes $(N, e, d) \leftarrow \mathsf{GenRSA}(1^k)$ and chooses random $z \leftarrow \mathbb{Z}_N^*$. It then outputs parameters (N, e, z) and trapdoor d. Let $D_N = \mathbb{Z}_N^*$.

- $\mathsf{Samp}(N)$ chooses a uniform element of $\mathbb{Z}_N^*$ in the trivial way.
- Given (N, e, z), define f_0 and f_1 as follows:

$$f_0(x) = x^e \bmod N \quad \text{and} \quad f_1(x) = z \cdot x^e \bmod N.$$

- Given (N, e, z) and the trapdoor information d, the inverses of f_0 and f_1 can be computed as follows: To compute $f_0^{-1}(y)$, simply compute $y^d \bmod N$. To compute $f_1^{-1}(y)$, simply compute $(y/z)^d \bmod N$.

The proof that the above is a clawfree trapdoor permutation follows the proof of Theorem 2.3, and is omitted. Observe that this construction is also doubly enhanced (once again, this comes "for free" due to the triviality of sampling from $\mathbb{Z}_N^*$).

In Chapter 4, we will introduce the (more recent) *strong* RSA assumption that offers additional flexibility as compared to the RSA assumption described here.

2.2.3 The Discrete Logarithm Assumption

An assumption of a different flavor is obtained by considering the *discrete logarithm problem*, which may be defined in any finite cyclic group $\mathbb{G}$. For the most part, we will consider only groups $\mathbb{G}$ of prime order q (though this is not required in any sense); such groups have the feature that every element in $\mathbb{G}$ other than the identity is a generator, and have several other advantages as well. Letting g be a generator of the group, and $h \in \mathbb{G}$ be arbitrary, define the discrete logarithm of h with respect to g (denoted $\log_g h$) as the smallest non-negative integer x such that $g^x = h$. (Note that we always have $\log_g h < q$.) The discrete logarithm problem is to compute x given g and a random group element h. We remark that it is easy to sample a random element $h \in \mathbb{G}$ by choosing a uniform integer $y \in \{0, \ldots, q-1\}$ and setting $h := g^y$.

For *certain* classes of groups, the discrete logarithm problem is believed to be hard. The problem is certainly not hard in *all* cyclic groups, and hardness depends to a great extent on the way elements of the group are represented. (This must be so, since all cyclic groups of the same order q are isomorphic yet the discrete logarithm problem is easy in the additive group $\mathbb{Z}_q$.)

To formally state the discrete logarithm assumption, we must consider an infinite sequence of groups as defined by an appropriate *group-generation algorithm*. Let $\mathcal{G}$ be a polynomial-time algorithm that, on input 1^k, outputs a (description of a) cyclic group $\mathbb{G}$, its order q (with q a k-bit integer), and a generator $g \in \mathbb{G}$. (Unless stated otherwise, we will also assume that q is prime except with negligible probability.) We also require that membership in $\mathbb{G}$ can be tested efficiently, and that the group operation in $\mathbb{G}$ can be computed efficiently (namely, in time polynomial in k). This implies that exponentiation in $\mathbb{G}$ can be performed efficiently as well.

Definition 2.10. The **discrete logarithm problem is hard relative to** $\mathcal{G}$ if for all PPT algorithms A, the following is negligible:

$$\Pr[(\mathbb{G}, q, g) \leftarrow \mathcal{G}(1^k); h \leftarrow \mathbb{G}; x \leftarrow A(1^k, \mathbb{G}, q, g, h) : g^x = h].$$

The discrete logarithm assumption is that there exists a $\mathscr{G}$ relative to which the discrete logarithm problem is hard.

We often abuse terminology and say that the discrete logarithm problem is hard for $\mathbb{G}$ when $\mathbb{G}$ is a group that is output by $\mathscr{G}$.

It is immediate that the discrete logarithm assumption implies the existence of a one-way function family; take $f_{\mathbb{G},q,g}(x)$ that outputs g^x. The functions in this family are one-to-one if we take the domain of $f_{\mathbb{G},q,g}(x)$ to be $\mathbb{Z}_q$. For certain groups $\mathbb{G}$ (for which the discrete logarithm problem is assumed to be hard), we can in fact obtain a one-way *permutation* family. One example, commonly used in cryptography, is given by groups of the form $\mathbb{Z}_p^*$ for p prime. (The order of this group is $q = p - 1$, and is not prime. The fact that groups of this form are cyclic is not obvious, but can be proved using basic field theory.) In this case, the "natural" mapping $f_{\mathbb{G},q,g} : \mathbb{Z}_{p-1} \to \mathbb{Z}_p^*$ is the one given above, where $f_{\mathbb{G},q,g}(x) = g^x \bmod p$. But by simply "shifting" the domain we get the function $f'_{\mathbb{G},q,g} : \mathbb{Z}_p^* \to \mathbb{Z}_p^*$ given by

$$f'_{\mathbb{G},q,g}(x) = g^{x-1} \bmod p.$$

We do not discuss other examples of groups for which the discrete logarithm problem is assumed to be hard. For the purposes of this book, abstracting the choice of group will be more convenient and suffices for our purposes.

The reader may be familiar with other assumptions related to the discrete logarithm assumption, most prominently the *computational* and *decisional* Diffie-Hellman assumptions. Interestingly, although these assumptions are extremely useful for the construction of efficient public-key encryption schemes, they have thus far had little application to the construction of efficient signature schemes. An exception is in the context of bilinear maps, discussed in Chapter 5, and we defer any discussion to there.

2.3 Hash Functions

Hash functions play a central role in the construction of secure signature schemes. Constructions of hash functions based on various assumptions are known, and we also have extremely efficient constructions (not based on any particular hard cryptographic problem) that one can reasonably assume to be secure. We will return to this point after defining the basic types of hash functions we will consider.

2.3.1 Definitions

A hash function is simply a function that *compresses* its input. Hash functions are used throughout computer science, but we will be interested in two types of *cryptographic* hash functions. Our hash functions will be *keyed*: that is, a function H is defined (this is sometimes also called a hash *family*); a key s is sampled uniformly at random and published; and we then look at the security of the keyed function H_s. A hash function H is said to be *collision-resistant* if, roughly speaking, it is hard to find distinct x, x' that "collide", that is, for which $H_s(x) = H_s(x')$. A hash function is called *universal one-way* if, informally, it is hard to find a collision for a *pre-specified* input x, chosen independently of the key s;

Definition 2.11. A **hash function** is a pair of probabilistic polynomial-time algorithms (Gen, H) such that:

- Gen is a probabilistic algorithm that on input 1^k outputs a key s. (We assume that 1^k is implicit in s.)
- There exists a polynomial ℓ such that H takes as input a key s and $x \in \{0,1\}^*$, and outputs a string $H_s(x) \in \{0,1\}^{\ell(k)}$. (Note that the running time of H is allowed to depend on $|x|$.)

If H_s is defined only for inputs $x \in \{0,1\}^{\ell'(k)}$, where $\ell'(k) > \ell(k)$ for all k, then we say that (Gen, H) is a **fixed-length hash function for inputs of length** ℓ'.

We now define the security properties stated informally earlier.

Definition 2.12. Hash function (Gen, H) is **collision-resistant** if the following is negligible for all PPT algorithms A:

$$\Pr\left[s \leftarrow \text{Gen}(1^k); (x, x') \leftarrow A(1^k, s) : x \neq x' \wedge H_s(x) = H_s(x')\right].$$

Definition 2.13. (Gen, H) is **universal one-way** if the following is negligible for all stateful PPT algorithms A:

$$\Pr\left[x \leftarrow A(1^k); s \leftarrow \text{Gen}(1^k); x' \leftarrow A(1^k, s) : x \neq x' \wedge H_s(x) = H_s(x')\right].$$

This is sometimes also called **second pre-image resistance** in the literature.

It is easy to see that collision-resistance implies universal one-wayness.

2.3.2 The Merkle-Damgård Transform

The Merkle-Damgård (MD) transform can be used to convert a fixed-length hash function (Gen, H') to a hash function (Gen, H) taking inputs of arbitrary length, while preserving collision-resistance. Besides being useful in its own right, the existence of the MD transform means that we can focus our attention on designing collision-resistant *compression functions* operating on short, fixed-length inputs, and then automatically convert such compression functions into full-fledged hash functions. The MD transform is also used extensively in practical constructions of hash functions.

Construction 2.1: The Merkle-Damgård transform

Let (Gen, H') be a fixed-length hash function for inputs of length $2\ell(k)$. (This assumption is for simplicity only.) Construct a hash function (Gen, H) as follows:

- The key-generation algorithm Gen is unchanged.
- H_s is defined for inputs of length at most $2^\ell - 1$. To compute $H_s(x)$ do:

  1. Set $L := |x|$ and $B := \lceil \frac{L}{\ell} \rceil$ (i.e., B is the number of blocks in x). Pad the input x with zeroes until its length is an integer multiple of ℓ, and parse the result as the sequence of ℓ-bit blocks $x_1, \ldots, x_B$. Set $x_{B+1} := L$, where L is encoded using exactly ℓ bits.
  2. Set $z_0 := 0^\ell$.
  3. For $i = 1, \ldots, B+1$, compute $z_i := H'_s(z_{i-1} \| x_i)$.
  4. Output z_{B+1}.

Theorem 2.5. *If (Gen, H') is collision-resistant, then so is (Gen, H).*

Proof (sketch). The proof follows easily from the observation that any collision in H_s yields a collision in H'_s. Let x and x' be two different strings, of lengths L and L' respectively, with $H_s(x) = H_s(x')$. Let $x_1, \ldots, x_B$ denote the padded version of x, and let $x'_1, \ldots, x'_{B'}$ be the result of padding x'. Recall further that $x_{B+1} = L$ and $x'_{B'+1} = L'$. Let z_i (resp, z'_i) be as in Construction 2.1. There are two cases to consider:

1. *Case 1: $L \neq L'$.* We have
$$H_s(x) = H'_s(z_B \| L) = H'_s(z'_{B'} \| L') = H_s(x').$$
   Since $z_B \| L \neq z'_{B'} \| L'$, this is a collision in H'_s.
2. *Case 2: $L = L'$.* In this case, note that $B = B'$ and $x_{B+1} = x'_{B+1}$. Since $x \neq x'$, there must exist an index i with $1 \leq i \leq B$ such that $x_i \neq x'_i$. Let $i^* \leq B+1$ be the *largest* index for which $z_{i^*-1} \| x_{i^*} \neq z'_{i^*-1} \| x'_{i^*}$. If $i^* = B+1$ then $z_B \| x_{B+1}$ and $z'_B \| x'_{B+1}$ are a collision in H'_s exactly as in the previous case. If $i^* \leq B$, then maximality of i^* implies $z_{i^*} = z'_{i^*}$, in which case $z_{i^*-1} \| x_{i^*}$ and $z'_{i^*-1} \| x'_{i^*}$ are a collision in H'_s.

The MD transform is only guaranteed to preserve collision-resistance; it is *not* guaranteed to preserve universal one-wayness. (A variant of the MD transform where an independent key is chosen for each iteration *does* preserve universal one-wayness, although this approach yields a hash function with a very long key. See further discussion at the end of Section 2.3.4.)

2.3.3 Constructing Collision-Resistant Hash Functions

In this section we describe constructions of (fixed-length) collision-resistant hash functions based on the number-theoretic assumptions introduced earlier in this chapter. We conclude with a brief discussion of hash functions used in practice.

Collision-resistant hash functions from clawfree permutations. We begin with a construction of collision-resistant hashing from any clawfree permutation family. Although not very practical, the construction provide a good illustration of how "strong" cryptographic primitives can be constructed from "weaker" building blocks. It also shows that the factoring and RSA assumptions imply the existence of collision-resistant hash functions.

Construction 2.2: Collision resistance from clawfree permutations

---

Let $(\mathsf{Gen}, \mathsf{Samp}, f_0, f_1)$ be a clawfree permutation family, and assume that if I is output by $\mathsf{Gen}(1^k)$ then elements of D_I can be described using ℓ bits. Define the fixed-length hash function (Gen_H, H) as follows:

- $\mathsf{Gen}_H(1^k)$ computes $I \leftarrow \mathsf{Gen}(1^k)$, chooses $r \leftarrow D_I$, and outputs the key $s = (I, r)$. In what follows, let $\ell = \ell(k)$ and write f_0, f_1 in place of $f_{I,0}, f_{I,1}$.
- $H_s(x)$, where $x \in \{0,1\}^{2\ell(k)}$, parses s as (I, r) and parses $x = x_1 \cdots x_{2\ell}$ where each x_i is a single bit. It then outputs $f_{x_{2\ell}}\left(f_{x_{2\ell-1}}\left(\cdots(f_{x_1}(r))\cdots\right)\right)$.

Theorem 2.6. *If $(\mathsf{Gen}, \mathsf{Samp}, f_0, f_1)$ is a clawfree permutation family, then hash function (Gen_H, H) from Construction 2.2 is collision-resistant.*

Proof (sketch). The proof follows easily from the observation that any collision in H_s yields a claw. For $x \in \{0,1\}^{2\ell}$ and $1 \le i \le 2\ell$ define

$$H_s^i = f_{x_i}\left(f_{x_{i-1}}\left(\cdots(f_{x_1}(r))\cdots\right)\right);$$

note that $H_s(x) = H_s^{2\ell}(x)$. Let $x = x_1 \cdots x_{2\ell}$ and $x' = x_1' \cdots x_{2\ell}'$ be two different strings with $H_s(x) = H_s(x')$, and let i denote the largest index such that $x_i \ne x_i'$ (so $x_j = x_j'$ for all $j > i$). Since f_0, f_1 are permutations,

$$H_s^{2\ell}(x) = H_s^{2\ell}(x') \;\Rightarrow\; H_s^i(x) = H_s^i(x') \;\Rightarrow\; f_{x_i}(H_s^{i-1}(x)) = f_{x_i'}(H_s^{i-1}(x'));$$

but then $H_s^{i-1}(x)$ and $H_s^{i-1}(x')$ are a claw.

Collision-resistant hash functions from the discrete logarithm assumption. In certain groups $\mathbb{G}$ (namely, where there is an efficiently computable bijection from $\mathbb{G}$ to $\mathbb{Z}_{|\mathbb{G}|}$), the discrete logarithm assumption implies a clawfree permutation family and thus a construction of a collision-resistant hash function as discussed above. We give a more direct and more efficient construction here, that additionally has the advantage of not working with arbitrary $\mathbb{G}$.

Let $\mathscr{G}$ be a group-generation algorithm, and assume that if $(\mathbb{G}, q, g) \leftarrow \mathscr{G}(1^k)$ then elements of $\mathbb{G}$ can be described using $k + O(1)$ bits. (This assumption is for simplicity only, and a generalization of the following construction works if this assumption does not hold.) Define a fixed-length hash function (Gen, H) as follows:

- $\text{Gen}(1^k)$ computes $(\mathbb{G}, q, g) \leftarrow \mathscr{G}(1^k)$ and chooses random $h \in \mathbb{G}$. It outputs the key $s = (\mathbb{G}, q, g, h)$.
- Let $s = (\mathbb{G}, q, g, h)$, and define $H_s : \mathbb{Z}_q \times \mathbb{Z}_q \to \mathbb{G}$ as

$$H_s(x, y) = g^x h^y.$$

If we want to view H_s as mapping bit-strings to bit-strings, note that H_s can handle inputs of length $2 \cdot (k - 1)$ (since any $(k - 1)$-bit integer can be viewed naturally as an element of $\mathbb{Z}_q$, using the fact that q is a k-bit integer), and the output of H_s can be encoded using $k + O(1)$ bits. Thus, for k large enough we have compression.

Theorem 2.7. *If the discrete logarithm problem is hard relative to $\mathscr{G}$, the above construction is collision resistant.*

Proof. Let A be a PPT collision-finding algorithm, and let

$$\varepsilon_A(k) \stackrel{\text{def}}{=} \Pr \left[\begin{array}{l} (\mathbb{G}, q, g, h) \leftarrow \text{Gen}(1^k); \\ (x, y, x', y') \leftarrow A(\mathbb{G}, q, g, h) \end{array} : (x, y) \neq (x', y') \wedge H_s(x, y) = H_s(x', y') \right]$$

(where $s = (\mathbb{G}, q, g, h)$) denote the probability with which A finds a collision. Construct the following PPT algorithm B solving the discrete logarithm problem relative to $\mathscr{G}$:

> **Algorithm B:**
> The algorithm is given $(\mathbb{G}, q, g, h)$ as input.
> Its goal is to compute $\log_g h$.
>
> - Run $A(\mathbb{G}, q, g, h)$ to obtain $x, y, x', y' \in \mathbb{Z}_q$.
> - If $y \neq y'$, output $(x - x') \cdot (y' - y)^{-1} \bmod q$. (Any $(y' - y) \neq 0$ has an inverse modulo q since q is prime.)

First note that A's input when run as a sub-routine by B is distributed identically to the keys output by Gen (this is true because B's input has h chosen uniformly from $\mathbb{G}$). Thus, A returns a collision with probability exactly $\varepsilon_A(k)$. We complete the proof of the theorem by showing that B outputs the correct answer $\log_g h$ whenever A finds a collision. To see this, note that

$$g^x h^y = g^{x'} h^{y'} \implies g^{x - x'} = h^{y' - y},$$

and if the above holds and furthermore $(x,y) \neq (x',y')$ then it must be the case that $y' - y \neq 0$. So $g^{(x-x')/(y'-y)} = h$, and this is exactly what is output by B.

Dedicated collision-resistant hash functions. We have seen that collision-resistant hash functions can be constructed based on a variety of number-theoretic assumptions. Yet these constructions are rather inefficient. In practice, dedicated constructions of (conjectured) collision-resistant hash functions are used that are orders of magnitude faster. These functions are generally *unkeyed* and have fixed length outputs. For these reasons, they cannot be said to satisfy asymptotic notions of security; nevertheless, appropriate *concrete* notions of security can be defined. Notable examples of hash functions in widespread use as of the time of this writing include SHA-1 (which hashes arbitrary-length inputs to 160-bit outputs) and SHA-256 (which hashes arbitrary-length inputs to 256-bit outputs).

2.3.4 Constructing Universal One-Way Hash Functions

Collision-resistance is a strong requirement. From a theoretical point of view, we currently know how to construct collision-resistant hash functions *only* from concrete, number-theoretic assumptions; constructions based on generic assumptions such as trapdoor permutations are not known. (Moreover, there is evidence that such constructions are impossible.) From a practical point of view, recent years have seen tremendous progress developing methods to attack hash functions. A prime example is the hash function MD5. For well over a decade MD5 was considered to be collision-resistant for all practical purposes. In 2005, however, Chinese cryptanalysts discovered a new technique for finding collisions in MD5. The attacks only got better, to the point where collisions in MD5 can now be found in minutes, and *structured* collisions in MD5 (i.e., colliding inputs x,x' that each satisfy certain formatting requirements) can now easily be found as well.

It is thus useful, when possible, to rely on the weaker assumption of universal one-wayness. Doing so potentially allows constructions based on weaker assumptions, and only makes it harder for an adversary to attack a deployed scheme. (In particular, MD5 is still considered to be universal one-way for the time being.) As an example of the former, we show here a construction of a (fixed-length) universal one-way hash function from any one-way permutation. (The construction can be easily modified to work with families of one-way permutations over bit-strings as well.) It is known that universal one-way hash functions can be constructed based on the (minimal) assumption of one-way *functions*; this construction is quite complex, unfortunately, and is beyond the scope of this book.

The construction of a universal one-way hash function from one-way permutations will use a particular pairwise-independent function family we now introduce. Let $\mathbb{F}_{2^k}$ denote the field with 2^k elements, and note that there is a natural correspondence between elements in $\mathbb{F}_{2^k}$ and k-bit strings. Define the function family

$$\mathscr{H}_k \stackrel{\text{def}}{=} \{h_{a,b} : \mathbb{F}_{2^k} \to \{0,1\}^{k-1} \mid b \in \mathbb{F}_{2^k}, a \in \mathbb{F}_{2^k} \setminus \{0\}\},$$

as

$$h_{a,b}(x) = \text{chop}(ax+b),$$

where chop simply removes the final bit of its input. We will use the following properties of $\mathcal{H}_k$:

Lemma 2.4. *For every* $b \in \mathbb{F}_{2^k}$, $a \in \mathbb{F}_{2^k} \setminus \{0\}$, *the function* $h_{a,b}$ *is two-to-one.*

Proof. Fix a,b and any $z \in \{0,1\}^{k-1}$. Let $z_0 = z\|0$ and $z_1 = z\|1$. The equation $ax+b = z_0$ has the unique solution $x = a^{-1} \cdot (z_0 - b)$ in $\mathbb{F}_{2^k}$ (using $a \neq 0$), and similarly for the equation $ax+b = z_1$.

Lemma 2.5. *Fix arbitrary* $y \in \mathbb{F}_{2^k}$, *and consider choosing* a,b *in the following way:*

1. *Choose uniform* $y' \in \mathbb{F}_{2^k} \setminus \{y\}$, *uniform* $z \in \{0,1\}^{k-1}$, *and uniform* $c \in \{0,1\}$. *Set* $z' = z\|c$ *and* $\bar{z}' = z\|\bar{c}$, *and view* $z', \bar{z}'$ *as elements of* $\mathbb{F}_{2^k}$.
2. *Solve for* a,b *in the following system of equations:*

$$ay+b = z' \tag{2.2}$$
$$ay'+b = \bar{z}'. \tag{2.3}$$

Then the distribution induced on (a,b) *is uniform over* $(\mathbb{F}_{2^k} \setminus \{0\}) \times \mathbb{F}_{2^k}$.

Proof. Note first that y', z, c determine a,b uniquely, and that $a \neq 0$ always. Now fix $a \in \mathbb{F}_{2^k} \setminus \{0\}$ and $b \in \mathbb{F}_{2^k}$ and let us see how many choices of y', z, c result in this (a,b) being chosen. Since a, y, b are now fixed, there is a unique choice of z, c satisfying Equation (2.2). Given this, $y' = (\bar{z}' - b) \cdot a^{-1} \neq y$ is the unique value satisfying Equation (2.3). We thus see that each pair (a,b) is selected with probability

$$\frac{1}{|\mathbb{F}_{2^k} \setminus \{0\}| \times |\{0,1\}^{k-1}| \times |\{0,1\}|} = \frac{1}{|\mathbb{F}_{2^k} \setminus \{0\}| \times |\mathbb{F}_{2^k}|},$$

and so the distribution of (a,b) is uniform over the indicated sets.

Let $f : \{0,1\}^* \to \{0,1\}^*$ be a length-preserving bijection. Define (Gen, H) as follows:

- $\text{Gen}(1^k)$ chooses uniform $a \in \mathbb{F}_{2^k} \setminus \{0\}$ and $b \in \mathbb{F}_{2^k}$, and outputs the key (a,b).
- $H_{a,b}(x)$ outputs $h_{a,b}(f(x))$.

Note that $H_{a,b}$ is two-to-one for every a,b; this follows from Lemma 2.4 and the fact that f is one-to-one.

Theorem 2.8. *If* f *is a one-way permutation, then the above hash function is universal one-way.*

Proof. Let A be a PPT collision-finding algorithm in the sense of Definition 2.13. Define

$$\varepsilon_A(k)$$
$$\stackrel{\text{def}}{=} \Pr\left[x \leftarrow A(1^k); (a,b) \leftarrow \text{Gen}(1^k); x' \leftarrow A(a,b) : x \neq x' \wedge H_{a,b}(x) = H_{a,b}(x')\right].$$

Construct the following PPT algorithm B inverting f:

> **Algorithm B:**
> The algorithm is given $y' \in \{0,1\}^k$ as input.
> Its goal is to compute $x' \in \{0,1\}^k$ with $f(x') = y'$.
>
> - Run $A(1^k)$ to obtain x; set $y = f(x)$. If $y' = y$ then output x and stop.
> - Otherwise choose random z, c and then compute a, b as in Lemma 2.5. Run $A(a,b)$ to obtain x'. Output x'.

If $y' = y$ then B clearly outputs an inverse of y. Conditioned on the event that this does not occur, y' is uniform in $\mathbb{F}_{2^k} \setminus \{y\}$. (This follows because B's input y' is computed as $y' = f(x')$ for uniform x', and f is a permutation.) It follows from Lemma 2.5 that a, b are distributed identically to the output of Gen, and so A finds a collision with probability exactly $\varepsilon_A(k)$ in this case.

By construction of B,

$$H_{a,b}(x) = \text{chop}(a \cdot f(x) + b) = \text{chop}(a \cdot y + b) = \text{chop}(z\|c) = z.$$

Since $H_{a,b}$ is two-to-one and

$$H_{a,b}(f^{-1}(y')) = \text{chop}(a \cdot y' + b) = \text{chop}(z\|\bar{c}) = z$$

collides with x, it follows that $f^{-1}(y')$ is the *only* input that collides with x, and so B outputs the correct result $f^{-1}(y')$ whenever A finds a collision. The theorem follows.

Improving the compression. The construction above only compresses its input by a single bit. Nevertheless, this suffices for constructing a universal one-way hash function mapping $p(k)$-bit inputs to $(k-1)$-bit outputs (for any desired polynomial p) using a variant of the Merkle-Damgård transform, with the difference being that *independent* keys must be used in each iteration. This gives a universal one-way hash function where the key size grows linearly in the input length and, in fact, is longer than the input. (This is no problem as far as the definition of universal one-wayness is concerned, but causes difficulty in some applications as will become clear in Chapter 3.) Transformations that improve the compression using smaller keys are also known; these can be used to construct universal one-way hash functions handling inputs of unbounded length. The details are beyond the scope of this book. Furthermore, as noted earlier, universal one-way hash functions can be constructed from one-way *functions*. For completeness we record the following:

Theorem 2.9. *Assuming the existence of one-way functions, there exist universal one-way hash functions (for arbitrary-length inputs).*

2.4 Applications of Hash Functions to Signature Schemes

We wrap up this chapter by showing how to use cryptographic hash functions to improve the parameters of digital signature schemes. We first show how to increase the message length of a signature scheme: specifically, we show how to convert a signature scheme capable of signing k-bit messages into one that can sign messages or arbitrary length. (We have already shown such a construction in Section 1.9, but the one given here is much more efficient.) This technique is used extensively in practice to enable signing large files. We then show how to decrease the size of public keys (at the expense of increasing the signature length), constructing in particular a one-time signature scheme that can sign messages twice as long as its own public key. This will form a crucial ingredient in our construction (in the following chapter) of an existentially unforgeable signature scheme from any one-way function.

2.4.1 Increasing the Message Length

Given a signature scheme for "short" messages, a natural way of handling longer messages is to *hash* all messages before signing them. We show how this can be implemented using both collision-resistant and universal one-way hash functions.

Using collision-resistant hash functions. When using collision-resistant hash functions, the above idea is simple to implement.

Construction 2.3: Increasing the message length using collision resistance

---

Let $\Pi = (\text{Gen}, \text{Sign}, \text{Vrfy})$ be a signature scheme for k-bit messages, and let (Gen_H, H) be a hash function mapping $p(k)$-bit inputs to k-bit outputs. Construct signature scheme $\Pi' = (\text{Gen}', \text{Sign}', \text{Vrfy}')$ for $p(k)$-bit messages as follows:

Key generation: $\text{Gen}'(1^k)$ computes $(pk, sk) \leftarrow \text{Gen}(1^k)$ and $s \leftarrow \text{Gen}_H(1^k)$. The public key is (pk, s) and the secret key is (sk, s).

Signature generation: Algorithm $\text{Sign}'_{sk,s}(m)$ outputs $\text{Sign}_{sk}(H_s(m))$.

Signature verification: Algorithm $\text{Vrfy}'_{pk,s}(m, \sigma)$ outputs $\text{Vrfy}_{pk}(H_s(m), \sigma)$.

Theorem 2.10. *If Π is existentially unforgeable (resp., strongly unforgeable) under an adaptive chosen-message attack and (Gen_H, H) is collision-resistant, then Π' is existentially unforgeable (resp., strongly unforgeable) under an adaptive chosen-message attack.*

Proof (sketch). The proof is quite straightforward, and so we merely provide a sketch for the case of existential unforgeability. Let $m_1, \ldots, m_\ell$ denote the messages submitted by an adversary A to the signing oracle $\text{Sign}'_{sk,s}(\cdot)$, and let (m, σ) denote a purported forgery output by A. There are two possibilities: either $H_s(m) = H_s(m_i)$ for some $i \in \{1, \ldots, \ell\}$ or not. If so, then A has found a collision in H_s (something that, by assumption on (Gen_H, H), occurs with only negligible probability). If not, then $H_s(m)$ is a k-bit string different from all k-bit strings $\{H_s(m_1), \ldots, H_s(m_\ell)\}$ that were signed using scheme Π. But then A has in fact output a valid forgery for Π (something that, by assumption on Π, occurs with only negligible probability).

Using universal one-way hash functions. In order to use universal one-way hash functions, we have to work a little harder. Note first that Theorem 2.10 is no longer guaranteed to hold if (Gen_H, H) is only universal one-way: In that case an adversary who observes s — which is included in the public key — might be able to find two different messages m, m' hashing to the same value, and then forge a signature on m' after requesting a signature on m. However, we *can* claim the following weaker version of Theorem 2.10:

Theorem 2.11. *If Π is existentially unforgeable (resp., strongly unforgeable) under a known-message attack and (Gen_H, H) is universal one-way, than Π' (as in Construction 2.3) is existentially unforgeable (resp., strongly unforgeable) under a known-message attack.*

Proof (sketch). The proof uses exactly the same ideas as in the proof of Theorem 2.10, the key difference being that in a known-message attack on Π' the adversary must "commit" to its messages $m_1, \ldots, m_\ell$ *before* it sees the public key (and so, in particular, before it sees the key s used for the hash function). Thus, if the adversary were able to output a forgery (m, σ) with $H_s(m) = H_s(m_i)$ for some i, this would violate the assumed universal one-wayness of (Gen_H, H).

The above already suffices to increase the message length for signature schemes secure against an adaptive chosen-message attack: given an existentially unforgeable signature scheme Π (which is in particular also secure against known-message attacks) for k-bit messages, apply the above theorem to obtain scheme Π' for arbitrary-length messages that is secure against *known* message attacks (and hence also against random-message attacks); then apply either of Theorems 1.1 or 1.2. As we shall see, a direct construction with better efficiency is possible.

The problem with Construction 2.3 (when a universal one-way hash function is used in place of a collision-resistant hash function) is that the adversary may select messages to be signed in a manner that depends on the hash key s included as part of the public key. We would like to prevent this, and "force" the adversary to choose the messages submitted to the signing oracle independently of the hash key. We can accomplish this by choosing the hash key "on the fly" as part of signature generation. Specifically, consider the following construction starting with an existentially unforgeable signature scheme $\Pi = (\text{Gen}, \text{Sign}, \text{Vrfy})$ and universal one-way hash function (Gen_H, H):

- Key generation is unchanged; i.e., compute $(pk, sk) \leftarrow \text{Gen}(1^k)$.

- To sign a message m using secret key sk, compute $s \leftarrow \mathsf{Gen}_H(1^k)$ and output the signature

$$(s, \mathsf{Sign}_{sk}(s|H_s(m)))\,.$$

  We stress that a *fresh* key s is computed for every message signed.
- To verify the signature (s, σ) on a message m with respect to a public key pk, output 1 iff $\mathsf{Vrfy}_{pk}(s|H_s(m), \sigma) \overset{?}{=} 1$.

The above construction is existentially unforgeable. To see this, let $m_1, \dots, m_\ell$ denote the messages submitted by the adversary A to its signing oracle, and let $(s_1, \sigma_1), \dots, (s_\ell, \sigma_\ell)$ be the signatures returned. Say A outputs forgery $(m, (s, \sigma))$ with $m \notin \{m_1, \dots, m_\ell\}$. Arguing as in the proof of Theorem 2.10, if $(s, H_s(m)) \neq (s_i, H_{s_i}(m_i))$ for all i then A has, in fact, generated a forgery in the original scheme Π (something that is assumed to occur with only negligible probability). On the other hand, if $(s, H_s(m)) = (s_i, H_{s_i}(m_i))$ but $m \neq m_i$ (making the simplifying, but inessential, assumption that the $\{s_i\}$ are distinct), then A has violated the assumed universal one-wayness of (Gen_H, H). (The key point being that the adversary chose m_i *before* it knew the hash key s_i.)

The main problem with this transformation is that the hash key *itself* is signed (along with the hashed message) by the underlying scheme, yet many theoretical constructions of universal one-way hash functions use rather long keys. In particular, when the hash key is longer than the input length of the hash function (as was the case for the construction described at the end of Section 2.3.4) the transformation is of no use. Instead, Construction 2.4 — which can be viewed as following the same paradigm used in Construction 1.2 — can be utilized.

Construction 2.4: Increasing the message length using universal one-wayness

---

Let $\Pi = (\mathsf{Gen}, \mathsf{Sign}, \mathsf{Vrfy})$ be a signature scheme for $2k$-bit messages, (Gen_H, H) be a hash function mapping $p(k)$-bit inputs to k-bit outputs and having keys of length $h(k)$, and (Gen'_H, H') be a hash function mapping $h(k)$-bit inputs to k-bit outputs. Construct signature scheme $\Pi' = (\mathsf{Gen}', \mathsf{Sign}', \mathsf{Vrfy}')$ for $p(k)$-bit messages as follows:

Key generation: $\mathsf{Gen}'(1^k)$ computes $(pk, sk) \leftarrow \mathsf{Gen}(1^k)$ and $s' \leftarrow \mathsf{Gen}'_H(1^k)$. The public key is (pk, s') and the secret key is (sk, s').

Signature generation: $\mathsf{Sign}'_{sk, s'}(m)$ computes $s \leftarrow \mathsf{Gen}_H(1^k)$ and outputs

$$\left(s, \mathsf{Sign}_{sk}(H'_{s'}(s) \,|\, H_s(m))\right).$$

Once again, we stress that a fresh key s is chosen for each message signed.

Signature verification: $\mathsf{Vrfy}'_{pk, s'}(m, (s, \sigma))$ outputs $\mathsf{Vrfy}_{pk}(H'_{s'}(s)|H_s(m), \sigma)$.

The reader is referred to [11] for a proof of the following:

Theorem 2.12. *If Π is existentially unforgeable (resp., strongly unforgeable) under an adaptive chosen-message attack and (Gen_H, H) and (Gen'_H, H') are both universal one-way, then Π' is existentially unforgeable (resp., strongly unforgeable) under an adaptive chosen-message attack.*

2.4.2 Reducing the Public-Key Length

As our next application of hash functions to signature schemes, we consider ways of shortening the public key. While these techniques are generally useful — in particular, they show that schemes with optimal public-key size are possible — our goal here is to use these techniques to construct a (one-time) signature scheme capable of signing messages *twice as long as* its own public key; this will be used when we construct existentially unforgeable signature schemes based on general assumptions in the next chapter.

The obvious way to decrease the public-key size is to simply hash the original public key. Formally, let $\Pi = (\mathsf{Gen}, \mathsf{Sign}, \mathsf{Vrfy})$ be an existentially unforgeable signature scheme having $q(k)$-bit public keys, and let (Gen_H, H) be a hash function mapping $q(k)$-bit inputs to k-bit outputs. Then the following scheme $\Pi' = (\mathsf{Gen}', \mathsf{Sign}', \mathsf{Vrfy}')$ has public keys of length k:

- $\mathsf{Gen}'(1^k)$ computes $(pk, sk) \leftarrow \mathsf{Gen}(1^k)$ and $s \leftarrow \mathsf{Gen}_H(1^k)$, and sets $pk' := H_s(pk)$. The public key is (s, pk') and the secret key is (pk, sk).
- $\mathsf{Sign}'_{pk,sk}(m)$ outputs $(pk, \mathsf{Sign}_{sk}(m))$.
- $\mathsf{Vrfy}_{s,pk'}(m, (pk, \sigma))$ outputs 1 iff (1) $H_s(pk) \overset{?}{=} pk'$ and (2) $\mathsf{Vrfy}_{pk}(m, \sigma) \overset{?}{=} 1$.

It is not difficult to verify that Π' is existentially unforgeable if (Gen_H, H) is universal one-way. (Note that the hashed input pk is chosen independently of the hash key s.) But public keys in Π' have length $|pk'| + |s| = k + |s|$ bits, an improvement only if $|s| < q(k) - k$. While this bound on the length of the hash key s can be achieved fairly easily if we are willing to assume collision-resistance, the bound is more difficult to ensure based on universal one-wayness alone (cf. the discussion at the end of Section 2.3.4).

Fortunately, it is possible to guarantee a public key shorter than the message by running sufficiently many copies of the original signature scheme in parallel.

Let us first verify the claim regarding the lengths of the public key and the messages. Π' has public keys of size $h(k) + \ell \cdot k$ and can sign messages of length $3k \cdot \ell$. By our choice of ℓ we have

$$3k\ell > 2k\ell + k \cdot (2h(k)/k) = 2k\ell + 2h(k),$$

and so the messages that can be signed have length at least twice that of the public key. As for the security of the construction, we have:

Construction 2.5: A signature scheme for messages twice as long as public keys

---

Let $\Pi = (\text{Gen}, \text{Sign}, \text{Vrfy})$ be a signature scheme for $3k$-bit messages having $q(k)$-bit public keys, and let (Gen_H, H) be a hash function mapping $q(k)$-bit inputs to k-bit outputs and having keys of length $h(k)$. Choose $\ell(k) > 2h(k)/k$, and construct signature scheme $\Pi' = (\text{Gen}', \text{Sign}', \text{Vrfy}')$ as follows:

Key generation: $\text{Gen}'(1^k)$ does as follows:

1. Compute $s \leftarrow \text{Gen}_H(1^k)$.
2. For $i = 1$ to ℓ, compute $(pk_i, sk_i) \leftarrow \text{Gen}(1^k)$ and set $pk_i' := H_s(pk_i)$.

The public key is $(s, pk_1', \ldots, pk_\ell')$, and the secret key is $(pk_1, sk_1, \ldots, pk_\ell, sk_\ell)$.

Signature generation: $\text{Sign}'_{pk_1, sk_1, \ldots, pk_\ell, sk_\ell}(m)$ parses m as $m_1, \ldots, m_\ell$ with $|m_i| = 3k$ for all i. It then outputs the signature $\left(pk_1, \text{Sign}_{sk_1}(m_1), \ldots, pk_\ell, \text{Sign}_{sk_\ell}(m_\ell)\right)$.

Signature verification: $\text{Vrfy}'_{s, pk_1', \ldots, pk_\ell'}(m, (pk_1, \sigma_1, \ldots, pk_\ell, \sigma_\ell))$ parses the message m as $m_1, \ldots, m_\ell$ with $|m_i| = 3k$ for all i. It then outputs 1 iff for all i: (1) $H_s(pk_i) \overset{?}{=} pk_i'$, and (2) $\text{Vrfy}_{pk_i}(m_i, \sigma_i) \overset{?}{=} 1$.

Theorem 2.13. *If Π is existentially unforgeable (resp., strongly unforgeable) under a one-time chosen-message attack and (Gen_H, H) is universal one-way, then Π' is existentially unforgeable (resp., strongly unforgeable) under a one-time chosen-message attack.*

Proof (sketch). We treat the case of existential unforgeability; strong unforgeability can be proven similarly. Consider a PPT adversary A attacking Π' in a one-time chosen-message attack. Let $m' = m_1', \ldots, m_\ell'$ be the message whose signature is requested by A, and say A outputs the forged signature $(\widehat{pk}_1, \sigma_1, \ldots, \widehat{pk}_\ell, \sigma_\ell)$ on the message $m = m_1, \ldots, m_\ell \neq m'$ If $\widehat{pk}_i \neq pk_i$ for some i, then A has violated the assumed universal one-wayness of (Gen_H, H). Letting j be any index with $m_j \neq m_j'$, we thus have that σ_j is a forged signature on the message m_j with respect to scheme Π (and public key pk_j).

An alternate, somewhat easier proof of the above theorem relies on the construction of universal one-way hash functions with high compression and short keys. We have given the above proof in order to keep the exposition self-contained.

Construction 2.5 is not existentially unforgeable when an adversary can request signatures on more than one message (even if Π is). However, a variant of the construction — in which each block of a signed message is pre-pended with a random, message-specific identifier — *is* secure in that sense (when Π is). See Construction 1.4 for the general idea.

2.5 Further Reading

Goldreich's book [56] is a good source for further information regarding generic assumptions, while more details regarding the number-theoretic assumptions discussed here can be found in [72]. The notions of a one-way function and a trapdoor permutation originate in the work of Diffie and Hellman [40], though formal definitions appeared only much later. Clawfree trapdoor permutations were introduced by Goldwasser, Micali, and Rivest [61] in the course of constructing the first secure digital signature scheme, and that work also contains a construction of clawfree trapdoor permutations based on the hardness of factoring (that is slightly different from the one given here).

Rabin [97] was the first to propose a trapdoor function based on the hardness of factoring, and Williams [110] and Blum [16] suggested restricting N to a special form to obtain a trapdoor *permutation*. As mentioned previously, the RSA assumption is due to Rivest, Shamir, and Adleman [99]. A recent survey by Boneh [17] discusses various attacks on RSA and also covers known results on the relationship between the RSA and factoring assumptions. The discrete logarithm assumption (without the restriction to prime order groups) is due to Diffie and Hellman [40].

Collision-resistant hash functions were first formally defined by Damgård [37], and the construction of collision-resistant hash functions from clawfree permutations is from that work as well. The Merkle-Damgård transformation was introduced independently in [38, 81], and our treatment in Section 2.3.2 is adapted from [72, Section 4.6.4]. Universal one-way hash functions originated in the work of Naor and Yung [88], where the construction of universal one-way hash functions based on one-way permutations was given. Rompel [100] (see also [71]) showed that universal one-way hash functions could be built from any one-way function. See [72] for extensive further discussion about hash functions and their applications.

Techniques for increasing the compression of universal one-way hash functions with reduced key expansion can be found in [11, 105]. Construction 2.4 is due to [11].

Part II
Digital Signature Schemes without Random Oracles

Chapter 3
Constructions Based on General Assumptions

Our objective in this chapter is to present a construction of a digital signature scheme based on the minimal assumption (cf. Theorem 2.1) that one-way functions exist. Along the way we will see a relatively simple construction, due to Lamport, of a *one-time* signature scheme based on the same assumption. We warn the reader at the outset that efficiency will not be a consideration here; we aim instead for generality (first) and simplicity of exposition (second). Interestingly, although several improved constructions of one-time signatures from one-way functions or permutations are known, the construction of a CMA-secure signature described in this chapter is essentially the best known (from any of the general assumptions discussed in the previous chapter); improving the efficiency of this generic construction remains an interesting and important open question.

Determining the weakest possible assumptions on which signatures can be based is, of course, essential for a solid theoretical understanding of signature schemes. Beyond this, one might wonder whether there is any practical value in studying schemes based on general assumptions given that all known examples of, say, one-way functions have some additional algebraic structure. In fact, this is not quite true: it is reasonable to treat, e.g., the hash function SHA-1 as a one-way function[1], and so we can use SHA-1 to instantiate the one-way function in schemes based on this assumption. Given the vast efficiency advantage of SHA-1 relative to number-theoretic assumptions, this may yield a scheme that is actually *more* efficient (at least[2] in certain respects) than the "efficient" schemes we show in later chapters. This is particularly true with regard to the Lamport one-time signature scheme.

---

[1] Since SHA-1 is defined only for fixed output length, it cannot be a one-way function in the complexity-theoretic sense. Nevertheless we certainly expect SHA-1 to be "hard to invert" for algorithms running in any practical amount of time.

[2] All known signature schemes based on general assumptions have very long signatures (regardless of how the underlying assumption is instantiated). On the other hand, the computational efficiency of signing or verifying (when using a scheme based on one-way functions that is instantiated with SHA-1) may — depending on the message length and the security desired — be competitive with the efficiency of number-theoretic schemes.

J. Katz, *Digital Signatures*, DOI 10.1007/978-0-387-27712-7_3,
© Springer Science+Business Media, LLC 2010

Overview of the chapter. We begin by describing Lamport's construction of a one-time signature scheme based on any one-way function. We then observe, in a series of steps, that we can use a tree-based approach to build a full-fledged signature scheme starting from any one-time signature scheme *that can sign messages twice as long as its public key.* It is worth remarking that a scheme with this property is trivial to construct if we are willing to assume the existence of collision-resistant hash functions. If we want a construction based only on one-way functions, we have to work a bit harder. The Lamport scheme, in particular, does not have this property (indeed, the public key in the Lamport scheme is much longer than the messages that can be signed). Fortunately, a signature scheme satisfying the stated requirement (and based only on one-way functions) follows fairly easily from our work in the previous chapter.

3.1 Lamport's One-Time Signature Scheme

The basic idea of Lamport's signature scheme is simple, and we illustrate it first for the case of signing 3-bit messages. Let f be a one-way function. The public key consists of 6 elements $y_{1,0}, y_{1,1}, y_{2,0}, y_{2,1}, y_{3,0}, y_{3,1}$ in the range of f; the private key contains the corresponding pre-images $x_{1,0}, x_{1,1}, x_{2,0}, x_{2,1}, x_{3,0}, x_{3,1}$. These keys can be visualized as two-dimensional arrays:

$$pk = \begin{pmatrix} y_{1,0} & y_{2,0} & y_{3,0} \\ y_{1,1} & y_{2,1} & y_{3,1} \end{pmatrix} \quad sk = \begin{pmatrix} x_{1,0} & x_{2,0} & x_{3,0} \\ x_{1,1} & x_{2,1} & x_{3,1} \end{pmatrix}.$$

To sign a message $m = m_1 \| m_2 \| m_3$, where each m_i is a single bit, the signer releases the appropriate pre-image x_{i,m_i} for $1 \leq i \leq 3$; that is, the signature σ simply consists of the three values $(x_{1,m_1}, x_{2,m_2}, x_{3,m_3})$. Verification is carried out in the natural way: presented with the candidate signature (x_1, x_2, x_3) on the message $m = m_1 \cdot m_2 \cdot m_3$, the verifier accepts if and only if $f(x_i) \stackrel{?}{=} y_{i,m_i}$ for $1 \leq i \leq 3$. This is shown graphically in Figure 3.1. The Lamport scheme for arbitrary message length ℓ is described formally as Construction 3.1.

Signing $m = 011$:

$$sk = \begin{pmatrix} \boxed{x_{1,0}} & x_{2,0} & x_{3,0} \\ x_{1,1} & \boxed{x_{2,1}} & \boxed{x_{3,1}} \end{pmatrix} \Rightarrow \sigma = (x_{1,0}, x_{2,1}, x_{3,1})$$

Verifying for $m = 011$ and $\sigma = (x_1, x_2, x_3)$:

$$pk = \begin{pmatrix} \boxed{y_{1,0}} & y_{2,0} & y_{3,0} \\ y_{1,1} & \boxed{y_{2,1}} & \boxed{y_{3,1}} \end{pmatrix} \Bigg\} \Rightarrow \begin{array}{c} f(x_1) \stackrel{?}{=} y_{1,0} \\ f(x_2) \stackrel{?}{=} y_{2,1} \\ f(x_3) \stackrel{?}{=} y_{3,1} \end{array}$$

Fig. 3.1 The Lamport scheme used to sign the message $m = 011$.

Construction 3.1: The Lamport one-time signature scheme

Let f be a function, and let $\ell = \ell(k)$ denote the desired message length.

Key generation: Algorithm $\mathsf{Gen}(1^k)$ is defined as follows. For $i \in \{1,\ldots,\ell\}$, do:

1. Choose random $x_{i,0}, x_{i,1} \leftarrow \{0,1\}^k$.
2. Compute $y_{i,0} := f(x_{i,0})$ and $y_{i,1} := f(x_{i,1})$.

The public key pk and the private key sk are

$$pk := \begin{pmatrix} y_{1,0} & y_{2,0} & \cdots & y_{\ell,0} \\ y_{1,1} & y_{2,1} & \cdots & y_{\ell,1} \end{pmatrix} \quad sk := \begin{pmatrix} x_{1,0} & x_{2,0} & \cdots & x_{\ell,0} \\ x_{1,1} & x_{2,1} & \cdots & x_{\ell,1} \end{pmatrix}.$$

Signature generation: On input a private key sk as above and a message $m \in \{0,1\}^\ell$ with $m = m_1 \| \cdots \| m_\ell$, output the signature $(x_{1,m_1}, \ldots, x_{\ell,m_\ell})$.

Signature verification: On input a public key pk as above, a message $m \in \{0,1\}^\ell$ with $m = m_1 \| \cdots \| m_\ell$, and a signature $\sigma = (x_1,\ldots,x_\ell)$, output 1 if and only if it is the case that $f(x_i) = y_{i,m_i}$ for $1 \le i \le \ell$.

Theorem 3.1. *Let ℓ be any polynomial. If f is a one-way function, then Construction 3.1 is existentially unforgeable under a one-time chosen-message attack.*

Proof. We let $\ell = \ell(k)$ for the rest of the proof. As intuition for the security of the scheme, note that for an adversary given public key $pk = \begin{pmatrix} y_{1,0} & y_{2,0} & \cdots & y_{\ell,0} \\ y_{1,1} & y_{2,1} & \cdots & y_{\ell,1} \end{pmatrix}$, finding an x such that $f(x) = y_{i^*,b^*}$ for *any* (i^*,b^*) amounts to inverting f. So it will certainly be hard to compute a signature on any message m given only the public key. What about computing a signature on some message m after being given a signature on a different message m'? If $m' \ne m$ then there must be at least one position i^* on which these messages differ. Say $m_{i^*} = b^* \ne m'_{i^*}$. Then forging a signature on m requires, in particular, finding an x such that $f(x) = y_{i^*,b^*}$. But finding such an x does not become any easier even when given $\{x_{i,b}\}$ for *all* $(i,b) \ne (i^*,b^*)$ (since the values $\{x_{i,b}\}_{(i,b)\ne(i^*,b^*)}$ are all chosen independently of x_{i^*,b^*}).

We now turn this intuition into a formal proof. Let $\Pi = (\mathsf{Gen},\mathsf{Sign},\mathsf{Vrfy})$ denote the Lamport signature scheme. Let A be a probabilistic polynomial-time adversary, and denote by $(m,\sigma) \leftarrow \mathsf{Expt}_{A,\Pi}(1^k)$ the experiment

$$(pk,sk) \leftarrow \mathsf{Gen}(1^k); (m,\sigma) \leftarrow A^{\mathsf{Sign}_{sk}(\cdot)}(pk),$$

where A is allowed only a *single* query to its signing oracle. Letting m' denote the message that A queries to its signing oracle in a particular execution of $\mathsf{Expt}_{A,\Pi}(1^k)$ (we assume without loss of generality that there is always some such message), we then let Forge be the event that $\mathsf{Vrfy}_{pk}(m,\sigma) = 1$ and $m \ne m'$.

Define
$$\mathsf{Succ}_{A,\Pi}(k) \stackrel{\text{def}}{=} \Pr\left[(m,\sigma) \leftarrow \mathsf{Expt}_{A,\Pi}(1^k) : \mathsf{Forge}\right]$$

and note that this is exactly the success probability of A as defined in Definition 1.7. To prove the theorem we must show that $\mathsf{Succ}_{A,\Pi}(k)$ is negligible.

In a particular execution of $\mathsf{Expt}_{A,\Pi}(1^k)$ where Forge occurs, we say that A *output a forgery at* (i,b) if $b = m_i \neq m'_i$. In other words, this means that A has succeeded in forging a valid signature on a message m after being given a signature on some message $m' \neq m$, and moreover $m_i = b$ while $m'_i = 1 - b$. It is immediately clear that whenever Forge occurs, A outputs a forgery at *some* (i,b).

We now construct the following PPT algorithm $\mathscr{I}$ attempting to invert the one-way function f:

> **Algorithm $\mathscr{I}$:**
> The algorithm is given y and 1^k as input. It goal is to output x with $f(x) = y$.
>
> 1. Choose a random index $i^* \leftarrow \{1,\ldots,\ell\}$ and a random bit $b^* \leftarrow \{0,1\}$. Set $y_{i^*,b^*} := y$.
> 2. For all $i \in \{1,\ldots,\ell\}$ and $b \in \{0,1\}$ with $(i,b) \neq (i^*,b^*)$:
>   - Choose $x_{i,b} \leftarrow \{0,1\}^k$ and set $y_{i,b} := f(x_{i,b})$.
> 3. Run A on input $pk := \begin{pmatrix} y_{1,0} & y_{2,0} & \cdots & y_{\ell,0} \\ y_{1,1} & y_{2,1} & \cdots & y_{\ell,1} \end{pmatrix}$.
> 4. When A requests a signature on the message m':
>   - If $m'_{i^*} = b^*$, stop.
>   - Otherwise, return the correct signature $\sigma = (x_{1,m'_1},\ldots,x_{\ell,m'_\ell})$.
> 5. When A outputs (m,σ) with $\sigma = (x_1,\ldots,x_p)$:
>   - If A output a forgery at (i^*,b^*), then output x_{i^*}.

Whenever A outputs a forgery at (i^*,b^*), algorithm $\mathscr{I}$ succeeds in inverting its given input y. We are interested in the probability that this occurs when the input to $\mathscr{I}$ is generated by choosing a random $x \leftarrow \{0,1\}^k$ and then setting $y := f(x)$ (cf. Definition 2.1). To analyze this probability it is useful to imagine a "mental experiment" in which $\mathscr{I}$ is given x at the outset, sets $x_{i^*,b^*} := x$, and then always returns a signature to A in step 4 (i.e., even if $m'_{i^*} = b^*$). It is not hard to see that the view of A being run as a subroutine by $\mathscr{I}$ in this mental experiment is distributed *identically* to the view of A in $\mathsf{Expt}_{A,\Pi}(1^k)$. Therefore, the probability that A outputs a forgery in step 5 is exactly $\mathsf{Succ}_{A,\Pi}(k)$. Because (i^*,b^*) was chosen at random at the beginning of the experiment, and the view of A is independent of this choice, the probability that A outputs a forgery at (i^*,b^*) — conditioned on the fact that A outputs a forgery — is at least $1/2\ell$. (This is because a signature forgery implies a forgery for at least one point (i,b). Since there are 2ℓ points, the probability of the forgery being at (i^*,b^*) is at least $1/2\ell$.) We conclude that, in this mental experiment, the probability that A outputs a forgery at (i^*,b^*) is at least $\mathsf{Succ}_{A,\Pi}(k)/2\ell$.

Returning to the real experiment involving $\mathscr{I}$ as initially described, the key observation is that *the probability that A outputs a forgery at (i^*,b^*) is unchanged.*

This is because the mental experiment and the real experiment coincide if A outputs a forgery at (i^*, b^*). That is, the experiments only differ if A requests a signature on a message m' with $m'_{i^*} = b^*$; but if this happens then it is impossible (by definition) for A to subsequently output a forgery at (i^*, b^*). So, in the real experiment, the probability that A outputs a forgery at (i^*, b^*) is still at least $\mathsf{Succ}_{A,\Pi}(k)/2\ell$. That is:

$$\Pr[x \leftarrow \{0,1\}^k; y := f(x); x' \leftarrow \mathscr{I}(1^k, y) : f(x') = y] \geq \mathsf{Succ}_{A,\Pi}(k)/2\ell.$$

Because f is a one-way function, we must have $\mathsf{Succ}_{A,\Pi}(k)/2\ell \leq \mathsf{negk}(k)$; since ℓ is polynomial, we conclude that $\mathsf{Succ}_{A,\Pi}(k)$ is negligible.

It is worth noting that the Lamport scheme is completely insecure if it is used to sign more than one message: an adversary who obtains signatures on both 0^ℓ and 1^ℓ (with respect to the same public key) learns the entire secret key!

Achieving strong unforgeability. An easy observation if that if the function f in Lamport's scheme is a one-way *permutation* (or generated from a family of one-way permutations in the natural way), then the scheme is *strongly unforgeable* under a one-time chosen-message attack. A variant of Lamport's scheme achieves strong unforgeability under the (minimal) assumption that one-way functions exist. The basic idea here is to instantiate the one-way function f in Construction 3.1 with a universal one-way hash function (that, in turn, can be constructed from any one-way function; see Section 2.3.4). In more detail, let (Gen_H, H) be a universal one-way hash function mapping $2k$-bit inputs to k-bit outputs. Key generation now works by choosing a random seed $s \leftarrow \mathsf{Gen}_H(1^k)$ and then proceeding as in Construction 3.1. I.e., for $i \in \{1, \ldots, \ell\}$, do:

1. Choose random $x_{i,0}, x_{i,1} \leftarrow \{0,1\}^{2k}$.
2. Compute $y_{i,0} := H_s(x_{i,0})$ and $y_{i,1} := H_s(x_{i,1})$.

The public key pk is

$$pk := \left(s, \begin{pmatrix} y_{1,0} & y_{2,0} & \cdots & y_{\ell,0} \\ y_{1,1} & y_{2,1} & \cdots & y_{\ell,1} \end{pmatrix}\right)$$

(note the inclusion of s), and the secret key is as before. Verification is done in the natural way. We have:

Theorem 3.2. *Let ℓ be any polynomial. If (Gen_H, H) is universal one-way, then the above construction is strongly unforgeable under a one-time chosen-message attack.*

Proof (sketch). The reduction is, in fact, easier than that used in the preceding proof. Construct an algorithm I as follows:

Algorithm I:

1. For all $i \in \{1, \ldots, \ell\}$ and $b \in \{0,1\}$, choose $x_{i,b} \leftarrow \{0,1\}^{2k}$.
2. Choose random $i^* \leftarrow \{1, \ldots, \ell\}$ and $b^* \leftarrow \{0,1\}$. Output x_{i^*, b^*} and receive in return a key s.

3. For all $i \in \{1, \dots, \ell\}$ and $b \in \{0, 1\}$, set $y_{i,b} := H_s(x_{i,b})$.

4. Run A on input $pk := \left(s, \begin{pmatrix} y_{1,0} & y_{2,0} & \cdots & y_{\ell,0} \\ y_{1,1} & y_{2,1} & \cdots & y_{\ell,1} \end{pmatrix} \right)$.

5. When A requests a signature on the message m', return the correct signature $\sigma = (x_{1,m_1'}, \dots, x_{\ell,m_\ell'})$.

6. When A outputs (m, σ) with $\sigma = (x_1, \dots, x_p)$:
   - If A output a forgery at (i^*, b^*), then output x_{i^*}.

Since I never need abort, it is not immediately clear that A outputs a forgery at (i^*, b^*) with probability at least $\mathsf{Succ}_{A,\Pi}(k)/2\ell$ (with $\mathsf{Succ}_{A,\Pi}(k)$ defined in the natural way). Now, however, an additional argument is needed in order to claim that I succeeds (cf. Definition 2.13) with overwhelming probability whenever A outputs a forgery at (i^*, b^*). (Indeed, it could be the case that $x_{i^*} = x_{i^*,b^*}$ in which case I does not succeed.) This follows from the facts that, with overwhelming probability, the number of pre-images of y_{i^*,b^*} with respect to H_s is exponential, and A has no information regarding which pre-image of y_{i^*,b^*} algorithm I chose.

3.2 Signatures from One-Time Signatures

Lamport's one-time signature scheme can be useful in its own right: although the public key (and signatures) are long, signing and verification can be very efficient when the one-way function is instantiated using, e.g., a cryptographic hash function. Lamport's scheme will also be a key component in our eventual construction of a CMA-secure signature scheme. Building to that result, we show here how to build a full-fledged signature scheme from any one-time signature scheme *that can sign messages twice as long as its public key*. Had Lamport's scheme satisfied this criterion, we would be done; alas, in Lamport's scheme the public key is much longer than the messages being signed. (E.g., if we take f to be length-preserving then a public key of length $2k\ell$ is needed to sign messages of length ℓ.) Nevertheless, the results of the preceding chapter already show how to construct a one-time signature scheme with the desired property, as we will later observe.

We build up to our final construction of this section in stages. In Section 3.2.1 we define the notion of a *stateful* signature scheme, where the signer updates its private key after each signature, and show how to construct a CMA-secure stateful signature scheme. In Section 3.2.2 we discuss a more efficient variant of this scheme that is still stateful. We then describe how this construction can be made stateless, as required by our original definition of a signature scheme.

3.2.1 "Chain-Based" Signatures

We first define the notion of a stateful signature scheme, where the signer may maintain some *state* that is updated after every signature is produced.

Definition 3.1. A **stateful signature scheme** consists of three efficient algorithms (Gen*, Sign*, Vrfy*) such that:

- The randomized **key-generation algorithm** Gen* takes as input the security parameter 1^k, and outputs (pk, sk, s_0) where pk is the **public key**, and sk is the **private key**, and s_0 is the **initial state**.
- The **signing algorithm** Sign* takes as input a secret key sk, the current state s_{i-1}, and a message m. It outputs a signature σ and an updated value s_i for the state.
- The deterministic **verification algorithm** Vrfy takes as input a public key pk, a message m, and a (purported) signature σ. It outputs a single bit b, with $b = 1$ signifying "accept" and $b = 0$ signifying "reject."

We impose the natural correctness condition: for every k, every (pk, sk, s_0) output by Gen*(1^k), and any sequence of messages $m_1, \ldots, m_\ell$, if we compute $(\sigma_i, s_i) \leftarrow$ Sign*$_{sk, s_{i-1}}(m_i)$ for $i \in \{1, \ldots, \ell\}$, then for every $i \in \{1, \ldots, \ell\}$ we have

$$\mathsf{Vrfy}^*_{pk}(m_i, \sigma_i) = 1.$$

We emphasize that the state is *not needed* in order to verify a signature. Signature schemes that do not maintain state (as per the standard definition) are called *stateless* to distinguish them from stateful schemes. Clearly, stateless schemes are preferable though stateful schemes can still potentially be useful depending on the context. In any case, as discussed earlier, our aim in introducing stateful signatures is simply to use them as a stepping stone to a full stateless construction.

Existential unforgeability under an adaptive chosen-message attack for the case of stateful signatures schemes is defined in a manner exactly analogous to Definition 1.6, with the only subtleties being that the signing oracle only returns the signature (and *not* the state), and that the signing oracle updates and maintains the state appropriately each time it is invoked.

We can easily construct a stateful "ℓ-time" signature scheme that can sign $\ell = \ell(k)$ messages for any polynomial ℓ. (The notion of security here would be analogous to the definition of one-time signatures given earlier; we do not give a formal definition since our discussion here is only informal.) Such a construction works by simply letting the public key consist of ℓ independently generated public keys for some one-time signature scheme Π, with the private key similarly constructed; i.e., set $pk := (pk_1, \ldots, pk_\ell)$ and $sk := (sk_1, \ldots, sk_\ell)$ where each (pk_i, sk_i) is an independently-generated key-pair for some one-time signature scheme. The state is just a counter initially set to 1. To sign a message m using the private key sk and current state $i \leq \ell$, simply compute $\sigma_i \leftarrow$ Sign$_{sk_i}(m)$ (that is, generate a one-time signature on m using the ith private key sk_i), output the signature (i, σ_i), and update the state to $i + 1$. In other words, the ith message is signed using sk_i. Verification of a signature (i, σ_i) on a message m can be done by checking whether σ_i is a valid signature on m with respect to pk_i.

Intuitively, this scheme is secure if used to sign ℓ messages because each private key is used to sign only a *single* message. (In fact, if Π is strongly unforgeable under a one-time chosen-message attack then the construction is strongly unforgeable under an ℓ-time chosen-message attack.) Since ℓ may be an arbitrary polynomial, why doesn't this give us the solution we are looking for? The main drawback is that the scheme requires the upper bound ℓ on the number of messages that can be signed *to be fixed in advance*, at the time of key generation. (In particular, the scheme does not satisfy Definition 1.2.) This is a potentially severe limitation since once the upper bound is reached a new public key would have to be generated and distributed. We would like instead to have a single, fixed scheme that can support signing an *unbounded* number of messages. Another drawback of the scheme is the fact that it is not very efficient, since the public and private keys have length that is linear in the total number of messages that can be signed. (We remark that more efficient constructions of ℓ-time signatures are possible, by direct modification of Lamport's scheme. Such scheme still have public and private keys whose length is linear in the number of messages to be signed and, in any case, the first drawback remains.)

Let $\Pi = (\mathsf{Gen}, \mathsf{Sign}, \mathsf{Vrfy})$ be a one-time signature scheme. In the scheme we have just described, the signer runs ℓ invocations of Gen to obtain public keys $pk_1, \ldots, pk_\ell$, and includes each of these in its actual public key pk. The signer is then restricted to signing at most ℓ messages. We can do better by using a "chain-based scheme" in which the signer generates and certifies additional public keys on-the-fly as needed.

Let us now assume that Π can sign messages twice as long as its public keys. For concreteness and without loss of generality, we assume that public keys have length k, and that messages of length $2k$ can be signed.

In the chain-based scheme, the public key consists of just a single public key pk_1 generated using Gen, and the private key contains the associated private key sk_1; the initial state is empty. To sign the first message $m_1 \in \{0,1\}^k$, the signer first generates a new key-pair (pk_2, sk_2) using Gen, and then signs both m_1 and pk_2 using sk_1 to obtain $\sigma_1 \leftarrow \mathsf{Sign}_{sk_1}(m_1 \| pk_2)$. The signature that is output includes both pk_2 and σ_1, and the signer adds $(m_1, pk_2, sk_2, \sigma_1)$ to its state. In the general case, when it comes time to sign the ith message the signer will have stored $\{(m_j, pk_{j+1}, sk_{j+1}, \sigma_j)\}_{j=1}^{i-1}$ as its state. To sign the ith message $m_i \in \{0,1\}^k$, the signer generates a new key-pair (pk_{i+1}, sk_{i+1}) using Gen, and then signs m_i and pk_{i+1} using sk_i to obtain a signature $\sigma_i \leftarrow \mathsf{Sign}_{sk_i}(m_i \| pk_{i+1})$. The actual signature that is output includes pk_{i+1}, σ_i, and also the values $\{m_j, pk_{j+1}, \sigma_j\}_{j=1}^{i-1}$. The signer then adds $(m_i, pk_{i+1}, sk_{i+1}, \sigma_i)$ to its state.

To verify a signature $(pk_{i+1}, \sigma_i, \{m_j, pk_{j+1}, \sigma_j\}_{j=1}^{i-1})$ on a message $m = m_i$ with respect to public key pk_1, the receiver verifies each link between the public key pk_j and the next public key pk_{j+1} in the chain, as well as the link between the last public key pk_{i+1} and m. That is, the verification procedure outputs 1 if and only if

$\mathsf{Vrfy}_{pk_j}(m_j \| pk_{j+1}, \sigma_j) \stackrel{?}{=} 1$ for all $j \in \{1, \ldots, i\}$. Observe that this verification begins from the public key pk_1 that was initially distributed.

It is not hard to be convinced — at least on an intuitive level — that the signature scheme thus constructed is existentially unforgeable under an adaptive chosen-message attack, regardless of how many messages are signed. (In fact, if Π is strongly unforgeable then so is the chain-based scheme constructed from Π.) Informally, this is once again due to the fact that each key-pair (pk_i, sk_i) is used to sign only a single "message" (where in this case the "message" is actually the concatenation $m_i \| pk_{i+1}$ of a message and a public key). Since we are going to prove the security of a more efficient scheme in the next section, we do not give a formal proof of security for the chain-based scheme here.

The chain-based signature scheme is a stateful signature scheme that is existentially unforgeable under an adaptive chosen-message attack. It has a number of disadvantages, though. For one, there is no immediate way to eliminate the state (recall that our ultimate goal is a stateless scheme satisfying Definition 1.2). It is also not very efficient, in that the signature length, size of the state, and verification time are all linear in the number of messages that have been signed. Finally, each signature reveals all previous messages that have been signed. While this does not technically violate any security requirement for signatures, it is easy to imagine that this may be undesirable in some contexts. We will eliminate all these advantages in the next section.

3.2.2 "Tree-Based" Signatures

The signer in the chain-based scheme of the previous section can be viewed as maintaining a *tree*, rooted at the public key pk_1, whose degree is 1 and whose depth is equal to the number of messages signed thus far. A natural way to improve the efficiency of this approach is to use a *binary* tree in which each node has degree 2. As before, a signature will correspond to a "certified" path in the tree from a leaf to the root; notice that as long as the tree has polynomial depth (even if it has exponential size!), verification can still be done in polynomial time.

Concretely, to sign messages of length k we will work with a binary tree of depth k having 2^k leaves. As before, the signer will add nodes to the tree "on-the-fly," as needed (and in particular this ensures that only polynomially many leaves are explicitly defined, while the rest remain implicit). In contrast to the chain-based scheme, though, only leaves (and not internal nodes) will be used to certify messages. Each leaf of the tree will correspond to one of the possible messages of length k.

In more detail, we imagine a binary tree of depth k where the root is labeled by ε (i.e., the empty string), and a node that is labeled with the binary string w of length less than k has left-child labeled $w0$ and right-child labeled $w1$. For every node w, we associate a pair of keys pk_w, sk_w from our one-time signature scheme Π. (We continue to assume that Π can sign messages up to twice as long as its public key.)

The public key of the root, pk_ε, is the actual public key of the signer. To sign a message $m \in \{0,1\}^k$, the signer carries out the following steps:

1. It first generates keys (as needed) for all nodes on the path from the root to the leaf labeled m. (Some of these public keys may have been generated in the process of signing previous messages; in this case the previous value — stored as part of the state — is used.)
2. Next, the signer "certifies" the path from the root to the leaf labeled m by computing a signature on $pk_{w0} \| pk_{w1}$, using private key sk_w, for each string w that is a proper prefix of m.
3. Finally, the signer "certifies" m itself by computing a signature on m using the private key sk_m.

The final signature on m consists of the signature on m with respect to pk_m, as well as all the information needed to verify the path from the leaf labeled m to the root. The signer also updates its state by storing all the key pairs generated as part of the signing process. A formal description of this scheme is given as Construction 3.2.

Notice that each of the underlying keys in this scheme is used to sign only a *single* "message": each key associated with an internal node signs a pair of public keys, and keys at the leaves are used to sign a single message.

Before proving security of this tree-based approach, let us note that it improves on the chain-based scheme in a number of respects. It still allows for signing an unbounded number of messages. In terms of efficiency, the signature length and verification time are now proportional to the message length k but are *independent* of the number of messages signed. (If some bound ℓ on the total number of messages to be signed were known, a modification of the scheme would have signature length and verification time $O(\log \ell)$.) The scheme is stateful, but we will see that this can be rectified after we prove the following result.

Theorem 3.3. *Let Π be a signature scheme that is existentially unforgeable (resp., strongly unforgeable) under a one-time chosen-message attack. Then Construction 3.2 is existentially unforgeable (resp., strongly unforgeable) under an adaptive chosen-message attack.*

Proof. We prove existential unforgeability, but the proof of strong unforgeability (assuming Π is strongly unforgeable) is essentially the same. Let Π^* denote Construction 3.2. Let A^* be a probabilistic polynomial time adversary, let $\ell^* = \ell^*(k)$ be a (polynomial) upper bound on the number of signing queries made by A^*, and set $\ell = \ell(k) \stackrel{\text{def}}{=} 2k\ell^*(k) + 1$. Note that ℓ upper-bounds the number of public keys from Π that are needed to generate ℓ^* signatures using Π^*. This is because each signature in π^* requires at most $2k$ new keys from Π (in the worst case), and there is one additional key from Π that is used as the actual public key pk_ε.

Let $\text{Expt}_{A^*, \Pi^*}(1^k)$ denote the experiment in which A^* interacts with Π^* exactly as in Definition 1.6 (with the only difference being that now the signing oracle maintains state). Let $\delta(k)$ denote the probability with which A^* outputs a valid forgery in $\text{Expt}_{A^*, \Pi^*}(1^k)$. Consider the following PPT adversary A attacking the one-time signature scheme Π:

Construction 3.2: A "tree-based" signature scheme

Let $\Pi = (\mathsf{Gen}, \mathsf{Sign}, \mathsf{Vrfy})$ be a signature scheme signing messages twice as long as its public key. For a binary string m, let $m|_i \stackrel{\text{def}}{=} m_1 \cdots m_i$ denote the i-bit prefix of m (with $m|_0 \stackrel{\text{def}}{=} \varepsilon$, the empty string).

Key generation: Algorithm $\mathsf{Gen}^*(1^k)$ is defined as follows. On input 1^k, compute $(pk_\varepsilon, sk_\varepsilon) \leftarrow \mathsf{Gen}(1^k)$ and output the public key pk_ε. The private key and initial state are sk_ε.

Signature generation: To sign a message $m \in \{0,1\}^k$ using the current state, algorithm Sign^* does the following:

1. For $i = 0$ to $k - 1$:

   - If $pk_{m|_i0}, pk_{m|_i1}$, and $\sigma_{m|_i}$ are not in the current state, compute them:

     $$(pk_{m|_i0}, sk_{m|_i0}) \leftarrow \mathsf{Gen}(1^k)$$
     $$(pk_{m|_i1}, sk_{m|_i1}) \leftarrow \mathsf{Gen}(1^k)$$
     $$\sigma_{m|_i} \leftarrow \mathsf{Sign}_{sk_{m|_i}}(pk_{m|_i0} \,\|\, pk_{m|_i1}),$$

   and then store all these computed values as part of the state.

2. If σ_m is not yet included in the state, compute $\sigma_m \leftarrow \mathsf{Sign}_{sk_m}(m)$ and store it as part of the state.

3. Output the signature $\left(\left\{ \sigma_{m|_i}, pk_{m|_i0}, pk_{m|_i1} \right\}_{i=0}^{k-1}, \sigma_m \right)$.

Signature verification: On input a public key pk_ε, a message $m \in \{0,1\}^k$, and signature $\left(\left\{ \sigma_{m|_i}, pk_{m|_i0}, pk_{m|_i1} \right\}_{i=0}^{k-1}, \sigma_m \right)$, output 1 if and only if:

1. $\mathsf{Vrfy}_{pk_{m|_i}}(pk_{m|_i0} \,\|\, pk_{m|_i1}, \sigma_{m|_i}) \stackrel{?}{=} 1$ for all $i \in \{0, \ldots, k-1\}$.

2. $\mathsf{Vrfy}_{pk_m}(m, \sigma_m) \stackrel{?}{=} 1$.

Adversary A:

A is given as input a public key pk (the security parameter k is implicit).

- Choose a random index $i^* \leftarrow \{1, \ldots, \ell\}$. Construct a list $pk^1, \ldots, pk^\ell$ of keys as follows:
  - Set $pk^{i^*} := pk$.
  - For $i \neq i^*$, compute $(pk^i, sk^i) \leftarrow \mathsf{Gen}(1^n)$.
- Run A^* on input the public key $pk_\varepsilon = pk^1$. When A^* requests a signature on a message m do:

1. For $i = 0$ to $k - 1$:
    - If the values $pk_{m|_i 0}, pk_{m|_i 1}$, and $\sigma_{m|_i}$ have not yet been defined, then set $pk_{m|_i 0}$ and $pk_{m|_i 1}$ equal to the next two unused public keys pk^j and pk^{j+1}, and compute a signature $\sigma_{m|_i}$ on $pk_{m|_i 0} \| pk_{m|_i 1}$ with respect to $pk_{m|_i}$.[3]

2. If σ_m is not yet defined, compute a signature σ_m on m with respect to pk_m (see footnote 3).

3. Give $\left(\left\{ \sigma_{m|_i}, pk_{m|_i 0}, pk_{m|_i 1} \right\}_{i=0}^{k-1}, \sigma_m \right)$ to A^*.

- Say A^* outputs a message m and a signature

$$\left(\left\{ \sigma'_{m|_i}, pk'_{m|_i 0}, pk'_{m|_i 1} \right\}_{i=0}^{k-1}, \sigma'_m \right).$$

If this is a valid signature on m, then:

Case 1: Say there exists a $j \in \{0, \dots, k-1\}$ for which $pk'_{m|_j 0} \neq pk_{m|_j 0}$ or $pk'_{m|_j 1} \neq pk_{m|_j 1}$; this includes the case when $pk_{m|_j 0}$ or $pk_{m|_j 1}$ were never defined by A. Take the minimal such j, and let j^* be such that $pk^{j^*} = pk_{m|_j} = pk'_{m|_j}$ (such a j^* exists by the minimality of j). If $j^* = i^*$, output $(pk'_{m|_j 0} \| pk'_{m|_j 1}, \sigma'_{m|_j})$.

Case 2: If case 1 does not hold, then $pk'_m = pk_m$. Let j^* be such that $pk^{j^*} = pk_m$. If $j^* = i^*$, output (m, σ'_m).

The final step of A may seem more complicated than it is due to the burdensome notation, but what is going on is rather simple: If A^* forges a signature (with respect to Π^*) on a message m, then this implies that A^* has effectively forged a signature (with respect to Π) somewhere along the path from the root to the leaf associated with m. The last step of A is simply to find the first such place where this occurs, and to identify the public key pk^{j^*} from its list of public keys $\{pk^1, \dots, pk^\ell\}$ where the forgery occurs. A succeeds in forging a signature (with respect to Π) if it happens to be the case that $pk^{j^*} = pk^{i^*} = pk$ (the public key A was given initially).

The view of A^* when run as a subroutine by A, when A is given pk generated by $\mathsf{Gen}(1^k)$, is distributed identically to the view of A^* in $\mathsf{Expt}_{A^*, \Pi^*}(1^k)$. Thus, the probability that A^* outputs a forgery when it is run as a subroutine by A is exactly $\delta(k)$. Assuming that A^* outputs a forgery, consider each of the two possible cases described above:

Case 1: Since i^* was chosen uniformly at random and is independent of the view of A^*, the probability that $j^* = i^*$ is exactly $1/\ell$. If $j^* = i^*$, then A requested a single signature on the message $pk_{m|_j 0} \| pk_{m|_j 1}$ with respect to the public key $pk = pk^{i^*} =$

---

[3] If $i \neq i^*$ then A can compute a signature with respect to pk^i by itself. For $i = i^*$, the algorithm A can obtain a (single) signature with respect to pk^{i^*} by making the appropriate query to its signing oracle. This is what is meant here.

$pk_{m|_j}$ that it was given. Moreover,

$$pk'_{m|_j0} \| pk'_{m|_j1} \neq pk_{m|_j0} \| pk_{m|_j1}$$

and yet $\sigma'_{m|_j}$ is a valid signature on $pk'_{m|_j0} \| pk'_{m|_j1}$ with respect to pk. Thus, A outputs a valid forgery in this case.

Case 2: Again, since i^* was chosen uniformly at random and is independent of the view of A^*, the probability that $j^* = i^*$ is exactly $1/\ell$. Assuming $j^* = i^*$, then A did not request any signatures with respect to the public key $pk = pk^{j^*} = pk_m$ and yet σ'_m is a valid signature on m with respect to pk. We see again that A outputs a valid forgery in this case.

That is, conditioned on A^* outputting a forgery (and regardless of which of the above cases occurs) A outputs a forgery with probability exactly $1/\ell$. This means that A succeeds in outputting a valid forgery with probability exactly $\delta(k)/\ell$. By the assumed security if Π and the fact that ℓ is polynomial, we conclude that $\delta(k)$ must be negligible.

Another approach. For completeness, we remark that a variant of the tree-based approach can be applied directly to any one-time signature scheme Π (i.e., even if Π cannot sign messages twice as long as its own public key). We describe this variant informally. The public key of Π^* contains pk_ε as in Construction 3.2 but now also include a key s for a universal one-way hash function. The same binary tree structure as in Construction 3.2 is used, except that now a node labeled w (associated with the key pair (pk_w, sk_w)) authenticates its two children (associated with the key pairs (pk_{w0}, sk_{w0}) and (pk_{w1}, sk_{w1}), respectively) by computing

$$\sigma_w \leftarrow \mathsf{Sign}_{sk_w} (H_s (pk_{w0} \| pk_{w1})).$$

(Messages are signed by leaves exactly as before, without any hashing.) The observation is that universal one-way hashing suffices here because all the public keys associated with internal nodes are outside the control of the attacker (i.e., they are chosen by the signer independent of the hash key s).

Reducing the tree depth. To sign messages of length ℓ, Construction 3.2 utilizes a tree of depth ℓ and associates each possible message $m \in \{0,1\}^\ell$ with the leaf labeled by m. (Of course, the underlying one-time signature scheme Π must be able to sign messages of length ℓ.) Since the signature time, verification time, and signature length all depend linearly on the tree depth, the efficiency of the scheme can be improved by decreasing the depth of the tree. A tree of depth $\omega(\log k)$ can be used (to sign messages of arbitrary length) in the following way: when signing a message, choose a *random* leaf; sign the message using the key associated with that leaf; and then authenticate the path from that leaf to the root exactly as in Construction 3.2. (It does not matter if different leaves end up being used to sign the same message. In particular, the signer need not store the leaves that are used to sign the different messages as part of its state [although it will continue to store the keys associated with each node in the tree, including the leaves].) If the same leaf ever gets used to sign two different messages, an adversary can forge a signature (using that same leaf); however, this occurs with only negligible probability since the number of leaves is super-polynomial. Besides taking this complication into account, the security proof is otherwise unchanged.

3.2.3 A Stateless Solution

In Construction 3.2, as described, the signer's state depends on the messages that have been signed. One could imagine, however, having the signer generate all necessary information (for all the nodes in the entire tree) *in advance*, at the time of key generation. (That is, at the time of key generation the signer could generate keys $\{(pk_w, sk_w)\}$ and signatures $\{\sigma_w\}$ for all binary strings w of length at most k.) If key generation were done in this way, the signer would not have to update its state at all; these values would instead all be stored as part of a (large) private key, and we would obtain a stateless scheme. The problem with this approach, of course, is that generating all these values at the time of key generation requires *exponential* time.

An alternative is to store some *randomness* that can be used to generate the values $\{(pk_w, sk_w)\}$ and $\{\sigma_w\}$, as needed, rather than storing the values themselves. That is, the signer could store a random string r_w for each w, and whenever the values pk_w, sk_w were needed the signer could compute $(pk_w, sk_w) := \mathsf{Gen}(1^k; r_w)$. (Similarly, if the signing procedure were probabilistic, the signer could store r'_w and then set $\sigma_w := \mathsf{Sign}_{sk_w}(pk_{w0} \| pk_{w1}; r'_w)$.) Generating and storing sufficiently many random strings at the time of key generation, however, would still require exponential time and space.

A simple modification of this alternative gives a polynomial-time solution. Instead of storing random r_w and r'_w as suggested above, the signer can store two keys k, k' for a *pseudorandom function F*. Roughly, a keyed function F is pseudorandom if the function $F_k(\cdot)$ (for random key k) is indistinguishable from a truly random function with the same domain and range. (The reader is referred to [72] for a formal definition of pseudorandom functions. This is the only place in the book where they are used.) When needed, the values pk_w, sk_w can now be generated by the following two-step process:

1. Compute $r_w := F_k(w)$.[4]
2. Compute $(pk_w, sk_w) := \mathsf{Gen}(1^n; r_w)$ (as before).

In addition, the key k' is used to generate the value r'_w that is used to compute the signature σ_w. This gives a *stateless* signature scheme in which key generation (as well as signing and verifying) can be carried out in polynomial time. Intuitively, this works because storing a random function is equivalent to storing all the r_w and r'_w values that are needed, and storing a pseudorandom function is "just as good".

---

[4] We assume F handles variable-length inputs, and has output length that is sufficiently long.

We leave it as an exercise to prove that this modified scheme remains existentially unforgeable under an adaptive chosen-message attack.

3.3 Signatures from One-Way Functions

We now have all the pieces in place to construct a CMA-secure signature scheme from any one-way function.

3.3.1 Putting the Pieces Together

The results of the previous section can be summarized as follows:

Theorem 3.4. *Assuming the existence of a signature scheme that is strongly unforgeable under a one-time chosen-message attack and can sign messages twice as long as its own public key, there exists a signature scheme that is strongly unforgeable under an adaptive chosen-message attack.*

Proof. A signature scheme as assumed in the theorem implies the existence of one-way functions (Theorem 2.1); these, in turn, are known to imply the existence of pseudorandom functions (see [64]). The discussion in the preceding section then gives the stated result.

Theorem 3.5. *Assuming the existence of one-way functions, there exists a signature scheme that is strongly unforgeable under an adaptive chosen-message attack.*

Proof. By the preceding theorem, it suffices to show how to construct a signature scheme Π that is strongly unforgeable under a one-time chosen-message attack and can sign messages twice as long as its own public key. Recall that the existence of one-way functions implies the existence of universal one-way hash functions (Theorem 2.9), and hence the existence of a signature scheme Π' that is strongly unforgeable under a one-time chosen-message attack (Theorem 3.2). Applying Construction 2.5 to Π' gives a signature scheme Π with the required properties.

3.3.2 Thoughts on the Construction

It is worth reflecting just how inefficient and complicated the resulting construction of a CMA-secure signature scheme is (when based directly on one-way functions, assuming no other primitives). Given a one-time signature scheme Π that can sign messages twice as long as its own public key, the stateful tree-based construction of a CMA-secure signature scheme Π^* described in Section 3.2.2 is conceptually simple but not very efficient, with the signing time, verification time, and signature length all depending linearly on the tree depth. (The tree depth is $O(k)$ as described in Section 3.2.2, but can be improved to $\omega(\log k)$ as discussed at the end of that section.) Making Π^* stateless requires a pseudorandom function: although block ciphers such as DES or AES (see [72]) provide efficient practical instantiations of pseudorandom functions, building a pseudorandom function from a one-way function is incredibly complex and inefficient.

Moreover, the underlying (one-time) scheme Π relies in an essential way on the construction of universal one-way hash functions from one-way functions — itself a complicated and inefficient process. Given such a hash function, building Π is not too complex (and, as noted following Theorem 3.5, can be further simplified) but is not very efficient, either. (The alternate approach discussed at the end of Section 3.2.2 still requires a universal one-way hash function.)

Some sort of tree-based approach seems essential to constructing signatures from one-way functions, although no proof of this is known. (One way to formalize the question would be to ask whether signature schemes built from one-way functions inherently have long signatures. In fact, a tree-based approach seems inherent even if we are willing to rely on stronger assumptions such as [clawfree] trapdoor permutations or collision-resistant hash functions.) More surprising, perhaps, is that all known constructions of (stateless) signature schemes from one-way functions rely on pseudorandom functions and universal one-way hash functions, even though there appears to be no inherent reason why this should be the case.

The above should only motivate the reader to either improve on the known constructions, or to show that further improvements are impossible!

3.4 Further Reading

We survey here the historical development of constructions of signature schemes based on progressively weaker assumptions, even though we have not covered all these early constructions in the text.

Lamport's one-time signature scheme was published in 1979 [76], though it was already described in [40]. Other, more efficient, constructions (that do not rely on specific number-theoretic assumptions but in some cases assume more than one-way functions) include those given in [79, 80, 45, 20, 15, 98, 85].

The first construction of a CMA-secure signature scheme was given by Goldwasser, Micali, and Rivest [61] based on the existence of clawfree trapdoor permutations. The Goldwasser-Micali-Rivest scheme was stateful, and Goldreich [55] (based on an idea credited to Levin) showed how to make it stateless using pseudorandom functions. Their scheme relies on a tree-based construction (similar, though not identical, to what is outlined in this chapter) whose genesis can be traced back to Merkle [80, 78, 79]. The idea for reducing the tree depth to $\omega(\log k)$ is also due to Goldreich [55].

Given the benefit of hindsight, it is interesting to observe that alternate — and, in some cases, simpler — constructions of *stateful* CMA-secure signature schemes could have been obtained at around the same time as the journal publication of [61]. (Note that the conference version of [61] appeared in 1984.) Goldwasser, Micali, and Yao [62] gave a KMA-secure scheme that could have been converted to a CMA-secure scheme using the techniques from Section 1.7.2. Their scheme is based on specific assumptions like RSA and factoring, but could have also been viewed as relying on clawfree trapdoor permutations. The tree-based approach described in Section 3.2.2 is based on ideas by Merkle [80] that go back to the late '70s; coupled with Lamport's scheme, this could have been used to construct signature schemes from collision-resistant hash functions via Construction 3.2. (Collision-resistant hash functions, in turn, can be constructed from clawfree permutations [37].) None of this is meant to take away anything from the incredible achievement of [61], which provides the first rigorous proof of security for any signature scheme construction.

Following the work of [61], the race was on to find the minimal assumptions under which signature schemes could be based. Bellare and Micali [5, 6] showed that trapdoor permutations suffice. Naor and Yung [88] introduced the concept of universal one-way hash functions; showed that these suffice for constructing signature schemes; and gave constructions of universal one-way hash functions from one-way permutations, this establishing that this assumption implies the existence of signature schemes. Subsequently, de Santis and Yung [39] showed that one-to-one one-way functions imply the existence of universal one-way hash functions. The question was finally settled by Rompel [100] (see also [71]), who showed that one-way functions suffice.

Apropos the discussion in Section 3.3.2, lower bounds on the (black-box) efficiency of *one-time* signature schemes based on general assumptions have been shown [52, 3].

Chapter 4
Signature Schemes Based on the (Strong) RSA Assumption

4.1 Introduction

The signature schemes described in the previous chapter have the advantage of being based on very weak cryptographic assumptions, but have the drawback of being incredibly inefficient. (Even the Lamport scheme, which could conceivably be used, has very large public keys and signatures.) It is natural to wonder whether relying on stronger, more specific assumptions might yield more efficient schemes. Unfortunately, progress in this direction has been limited: only a handful of schemes are known that are more efficient than the "generic" constructions of the previous chapter and, of these, even fewer are efficient enough to compete with the signature schemes currently used in practice.[1] In fact, and somewhat disappointingly, the *only* schemes we currently have that come close to the efficiency of signature schemes currently in use are based on relatively "new" cryptographic assumptions discussed in this and the following chapter. (Admittedly, this point is debatable and depends to some extent on what one takes as his measure of efficiency.)

In this chapter we present signature schemes based on the RSA assumption or a recent variant (discussed later) called the "strong" RSA assumption. In the following chapter we will introduce schemes that rely on bilinear maps.

4.1.1 Technical Preliminaries

We begin with some technical lemmas that will prove useful in the analysis of the schemes discussed in this chapter.

The RSA assumption (informally) is that given a modulus N (that is a product of two primes), a random element $y \in \mathbb{Z}_N^*$, and an exponent e that is relatively prime to $\phi(N)$, it is "hard" to compute the eth root of y (modulo N). Might it be easier to

---

[1] We stress that here we are referring to schemes that can be proven secure *without* resorting to the random oracle model. Part III of this book discusses schemes of the latter type.

compute the eth root of some *power* of y; i.e., to compute the eth root of $y^{e'} \bmod N$ for some e' of our choice? The following lemma shows that this is not the case as long as e and e' are relatively prime.[2] Specifically, the lemma states that if e, e' are relatively prime then an eth root of $y^{e'} \bmod N$ can be used to compute an eth root of y itself. (Note that the lemma does not require e or e' to be relatively prime to $\phi(N)$.)

Lemma 4.1. *Given N, elements $x, y \in \mathbb{Z}_N^*$, and integers e, e' for which it holds that $\gcd(|e|, |e'|) = 1$ and $x^e = y^{e'} \bmod N$, an eth root of y (modulo N) can be computed in polynomial time.*

Proof. Assume without loss of generality that e, e' are positive (if not, we can always re-write the equation by taking inverses). Applying the extended Euclidean algorithm to e, e' we can efficiently compute integers A, B satisfying

$$Ae + Be' = \gcd(e, e') = 1.$$

We claim that $y^A x^B \bmod N$ (which can be computed easily given what is known) is an eth root of y. Indeed, we have

$$\left(y^A x^B\right)^e \bmod N = y^{Ae}(x^e)^B \bmod N$$
$$= y^{Ae}(y^{e'})^B \bmod N$$
$$= y^{Ae+Be'} = y^1 = y \bmod N,$$

concluding the proof.

A nice consequence of the above is the observation, due to Shamir [102], that it does not become any easier to compute $y^{1/e} \bmod N$ even given $\{y^{1/e_i} \bmod N\}$ for some set of primes $\{e_i\}$ not containing e. More formally:

Lemma 4.2. *Say the RSA problem is hard relative to GenRSA, and let $\{e_i\}_{i=1}^{\ell}$ be a set of primes that does not contain the exponent e output by GenRSA and such that $\gcd(\phi(N), e_i) = 1$ for all N output by GenRSA and all i. Then the following is negligible for all PPT algorithms A:*

$$\Pr\left[\begin{array}{c} (N, e, d) \leftarrow \mathsf{GenRSA}(1^k); y \leftarrow \mathbb{Z}_N^*; \\ x \leftarrow A(1^k, N, e, y, \{e_i, y^{1/e_i}\}_{i=1}^{\ell}) \end{array} : x^e = y \bmod N \right]. \quad (4.1)$$

Proof. Given a PPT algorithm A, construct the following algorithm A' which attempts to solve the (standard) RSA problem:

> **Algorithm A':**
> The algorithm is given (N, e, y) as input.
> Its goal is to compute $y^{1/e} \bmod N$.
>
> • Set $\hat{e} := \prod_{i=1}^{\ell} e_i$.

---

[2] Note that it may be trivial to compute the eth root of $y^{e'}$ if $\gcd(e, e') \neq 1$. For example, if $e' = k \cdot e$ for an integer k then y^k is the eth root of $y^{e'}$.

- Set $\hat{e}_i := \hat{e}/e_i$ for $1 \le i \le \ell$. (Note $\hat{e}_i$ is an integer for all i.)
- Set $Y := y^{\hat{e}} \bmod N$ and $Y_i := y^{\hat{e}_i} \bmod N$ for $1 \le i \le \ell$.
- Run $A(1^k, N, e, Y, \{e_i, Y_i\}_{i=1}^{\ell})$ and obtain output X.
- If $X^e \stackrel{?}{=} Y = y^{\hat{e}} \bmod N$, then compute $x = y^{1/e} \bmod N$ using Lemma 4.1.

The key to the above algorithm is that the $\hat{e}_i$ are all integers, and so A' can efficiently compute the values

$$Y_i = y^{\hat{e}_i} = y^{\hat{e}/e_i} = Y^{1/e_i} \bmod N.$$

Thus, A' is a PPT algorithm. Furthermore, Lemma 4.1 does indeed apply since $\gcd(e, \hat{e}) = 1$ (this follows easily since the $\{e_i\}$ are prime and $e \notin \{e_i\}$). We conclude that A' successfully outputs the eth root of its given input value y whenever A successfully outputs the eth root of its input value Y. The theorem follows by noting that the distribution on the inputs given to A in the above algorithm is identical to the distribution on the inputs given to A in Equation (4.1): since $\gcd(\phi(N), \prod_i e_i) = 1$, the value $Y = y^{\hat{e}}$ is uniformly distributed in $\mathbb{Z}_N^*$ when y is uniformly distributed in $\mathbb{Z}_N^*$, and it is easy to see from what we have already said that the remainder of A's inputs have the correct distribution.

We conclude that A' computes the eth root of y with probability exactly that given by Equation (4.1). If the RSA problem is hard relative to GenRSA, then, Equation (4.1) must be negligible for any PPT algorithm A as claimed.

We do not directly use Lemma 4.2 in what follows, though we will often implicitly rely on the techniques used in its proof. To get a feeling for why the above lemma might be useful in our context, note that it suggests the following very simple signature scheme (which only supports a polynomial-size message space as described): the public key contains $(N, y, \{e_i\}_{i=1}^{\ell})$ (where the $\{e_i\}$ satisfy the conditions of the lemma above); a signature on the message i (for $1 \le i \le \ell$) is given by $y^{1/e_i} \bmod N$. Lemma 4.2 implies that it is hard to compute the signature on a message j even when given the signatures on all other messages $\{i\}_{i \ne j}$. (This can be turned into a formal proof.) We will see in Section 4.3.5 that an extension of this idea yields a relatively efficient scheme (which is, however, proved secure based on a stronger assumption than standard RSA).

Lemma 4.2 requires a set of (distinct) primes $\{e_i\}$ each of which is relatively prime to $\phi(N)$. In the applications that follow, it will be convenient to fix this set of primes independently of N (and, in particular, before N is even known). An easy way to achieve this is to choose the primes such that $e_i > \max\{p, q\}$, where p, q are the factors of N. This is easy to ensure (without knowledge of the factorization of N) by choosing e_i as a $(k+1)$-bit prime (recall that p and q, the factors of N, are k-bit integers when GenRSA is run on security parameter 1^k).

4.1.2 Outline of the Chapter

We present here a number of signature schemes based on the RSA and strong RSA assumptions. We begin by describing a signature scheme proposed by Dwork and Naor, as well as an improvement of this scheme due to Cramer and Damgård. Both of these schemes rely on a tree-based approach similar to that used in the previous chapter, with the efficiency improvements being due to the fact that, by relying on the RSA assumption, trees of larger degree (and hence lower depth) can be used. We then show a recent scheme by Hohenberger and Waters that can also be viewed as using a tree-based approach, but has the advantage of yielding shorter signatures. All these schemes are proven secure based on a variant of the standard RSA assumption (see the next section for details).

The RSA assumption states that, given N, e, and y, it is infeasible to compute $y^{1/e} \bmod N$. The *strong RSA* assumption (described formally in Section 4.3) is that, given N and y, it is infeasible to compute $(e, x^{1/e})$ for *any* $e \geq 2$, even if we allow the freedom to choose e. Cramer and Shoup have shown how the strong RSA assumption can be used to dramatically simplify and improve the efficiency of the Cramer-Damgård scheme; we discuss the resulting Cramer-Shoup signature scheme, as well as further improvements of this scheme due to Fischlin, in Sections 4.3.3 and 4.3.4. A different approach, but also relying on the strong RSA assumption, is taken in the Gennaro-Halevi-Rabin scheme shown in Section 4.3.5.

4.2 Signature Schemes Based on the RSA Assumption

In this section we will show three schemes based on the RSA assumption. Each of these schemes, as described, actually requires a variant of the RSA assumption that we formalize now.

Throughout this section, let GenModulus be a probabilistic polynomial-time algorithm that, on input 1^k, outputs a modulus N along with two (distinct) k-bit primes p, q with $N = pq$. Let GenPrime be a probabilistic polynomial-time algorithm that, on input 1^k, outputs a prime e of length at least k. We require that GenPrime generates primes "at random" in the sense that if we run GenPrime twice, the outputs of these two executions are equal with only negligible probability. (This immediately implies that even if we run GenPrime polynomially many times, the probability of getting a repeated output is negligible.) We highlight that GenPrime does not take N or the factors of N as input, but if e is a prime of length at least k (and N is a product of k-bit primes) then we are guaranteed that $\gcd(e, \phi(N)) = 1$.

We use the following variant of the RSA assumption in Sections 4.2.1 and 4.2.2:

Definition 4.1. The RSA problem is hard relative to GenModulus **and** GenPrime if for all polynomials p (with $p(k) \geq k$ for all k) and all PPT algorithms A, the following is negligible:

$$\Pr\left[\begin{array}{l}(N,p,q) \leftarrow \mathsf{GenModulus}(1^k); y \leftarrow \mathbb{Z}_N^*; \\ e \leftarrow \mathsf{GenPrime}(1^{p(k)}); x \leftarrow A(N,e,y)\end{array} : x^e = y \bmod N\right].$$

In other words, this definition requires that the RSA problem is hard relative to a "random" large public exponent e.

In Section 4.2.3 we use a slightly stronger version of the above:

Definition 4.2. The RSA problem is hard relative to $\mathsf{GenModulus}$ **and** $\mathsf{GenPrime}$, **even with public coins** if for all PPT algorithms A, the following is negligible:

$$\Pr\left[\begin{array}{l}(N,p,q) \leftarrow \mathsf{GenModulus}(1^k); y \leftarrow \mathbb{Z}_N^*; \\ \omega \leftarrow \{0,1\}^k; e \leftarrow \mathsf{GenPrime}(1^k; \omega); x \leftarrow A(N,e,y,\omega)\end{array} : x^e = y \bmod N\right].$$

In other words, this definition requires that the RSA problem is hard relative to a "random" large public exponent e, even if the random coins used to generate e are known.

4.2.1 The Dwork-Naor Scheme

Dwork and Naor [42] showed that by relying on *specific properties* of the RSA problem it is possible to construct signature schemes that are more efficient than what would be obtained by using the generic construction of the previous chapter. As in Construction 3.2, the signer in the Dwork-Naor scheme maintains a tree of vertices. Each vertex is here associated with a *label* that, for now, can be viewed as roughly analogous to the public key that is associated with each vertex in Construction 3.2. Continuing the analogy to Construction 3.2, here too the label of the root vertex is included as part of the public key; vertices authenticate their children; and messages are signed by associating them with leaves of the tree (although the exact way this is done is now slightly different). The main novelty of the Dwork-Naor scheme — and the key feature that makes it more efficient than a naive instantiation of Construction 3.2 — is that in the Dwork-Naor scheme a vertex can authenticate each of its children *independently* instead of having to authenticate its children all at once. Specifically, in the Dwork-Naor scheme a verifier can check that a particular node v (with some known label) is authenticated by its parent *without having to know the labels of any of v's siblings*. (In contrast, the verifier in Construction 3.2 needed to know the labels of all of v's siblings in order to verify authenticity of v.) As a consequence, the Dwork-Naor scheme can more readily use a tree of larger degree, and hence smaller depth, resulting in a more efficient construction.[3] We now describe in more detail exactly how this is accomplished.

Fix some integer $\ell \geq 2$ representing the degree of the tree that will be constructed as part of the scheme. The public key in the Dwork-Naor scheme includes a modulus

---

[3] Of course, Construction 3.2 could also have used a tree of larger degree (given a signature scheme capable of signing sufficiently long messages), but the reader can check that this will reduce the efficiency in that case.

N (a product of two k-bit primes) as well as a list $Y = \{y_1, \ldots, y_{2k}\}$ of elements from $\mathbb{Z}_N^*$ and a set of distinct primes $E = \{e_1, \ldots, e_\ell\}$ that are all relatively prime to $\phi(N)$. In the tree that will be constructed vertices will be labeled with strings of length $2k$; observe that elements of $\mathbb{Z}_N^*$ can be represented as strings of this length. The basic authentication step, which we now describe, enables a vertex with label L to authenticate its ith child ($1 \leq i \leq \ell$) having label L', and is defined as follows:

$$\mathsf{auth}(L \xrightarrow{i} L') \stackrel{\text{def}}{=} \left(L \cdot \prod_{j:[L']_j=1} y_j \right)^{1/e_i} \bmod N \qquad (4.2)$$

(in the above, $[L']_j$ refers to the jth bit of L'). This authentication information can be computed if the factorization of N is known. Note also that given a (presumably authenticated) parent vertex with label L, its candidate ith child with label L', and authentication information auth, it is easy to verify the authenticity of L' with respect to a known public key by checking whether

$$\mathsf{auth}^{e_i} \stackrel{?}{=} L \cdot \prod_{j:[L']_j=1} y_j \bmod N. \qquad (4.3)$$

With the above basic authentication step in place, designing a full-fledged signature scheme is relatively straightforward given the results of the previous chapter. The signer will maintain a tree of degree ℓ and depth d, where the root of the tree is assigned a label that is included as part of the public key. Messages will be associated with the leaves of this tree, meaning that the scheme can be used to sign at most[4] $B = \ell^d$ messages. (For a fixed value of B, larger values of ℓ translate to a larger public key but shorter signatures.) To sign a message $m \in \{0,1\}^{2k}$, the signer assigns the label m to the next unused leaf, and then authenticates the path from the root to this leaf. The verifier validates the claimed sequence of authenticators in the obvious way. Note that the scheme, as described, is stateful since the signer must keep track of both which leaves have been used as well as the labels assigned to internal nodes in the tree.[5]

The scheme is formally described as Construction 4.1.

Efficiency of the scheme can be improved by using a collision-resistant hash function H to map elements of $\mathbb{Z}_N^*$ to shorter strings. In this case, a basic authentication step would take the form

$$\mathsf{auth}(L \xrightarrow{i} L') \stackrel{\text{def}}{=} \left(L \cdot \prod_{j:[H(L')]_j=1} y_j \right)^{1/e_i} \bmod N.$$

---

[4] It is not difficult to extend the scheme so that an unbounded number of messages can be signed; for simplicity, however, the scheme is described with a fixed upper bound on the number of signatures to be issued.

[5] Although it is possible to avoid maintaining state using essentially the same techniques described in Section 3.2.3, doing so would reduce the efficiency of the scheme.

Construction 4.1: The Dwork-Naor scheme

Let GenModulus, GenPrime be as described in the text.

Key generation: On input security parameter 1^k, proceed as follows:

- Run $(N, p, q) \leftarrow$ GenModulus(1^k).
- For $i \in \{1, \ldots, \ell\}$, compute $e_i \leftarrow$ GenPrime(1^k). Set $E := \{e_1, \ldots, e_\ell\}$.
- For $i \in \{1, \ldots, 2k\}$, let y_i be a uniformly distributed element in $\mathbb{Z}_N^*$. Also choose $L_{v_0} \leftarrow \mathbb{Z}_N^*$. Set $Y := \{y_1, \ldots, y_{2k}\}$.
- The public key is (N, L_{v_0}, Y, E) and the secret key is p, q.

Signature generation: The signer implicitly holds a tree of depth d and out-degree ℓ whose root v_0 has label L_{v_0}. To generate a signature on a message $m \in \{0, 1\}^{2k}$, let v_d be the left-most leaf in the tree that has not yet been used. Assign label $L_{v_d} := m$ to this leaf. Let $(i_1, \ldots, i_d)$ denote the sequence of edges on the path from the root to this leaf (where each $i_1, \ldots, i_d$ lies in the range $\{1, \ldots, \ell\}$), and let $(v_1, \ldots, v_d)$ be the nodes on this path (not including the root). Let $(L_{v_1}, \ldots, L_{v_d})$ denote the labels of these nodes (if any of these nodes have not yet been assigned a label, they are now assigned a label chosen uniformly at random from $\mathbb{Z}_N^*$). The signature on m is then:

$$(i_1, \ldots, i_d), (L_{v_1}, \ldots, L_{v_{d-1}}), \mathsf{auth}(L_{v_0} \xrightarrow{i_1} L_{v_1}), \ldots, \mathsf{auth}(L_{v_{d-1}} \xrightarrow{i_d} L_{v_d}),$$

where the $\mathsf{auth}(L \xrightarrow{i} L')$ are computed as in Equation (4.2).

Signature verification: A signature

$$(i_1, \ldots, i_d), (L_{v_1}, \ldots, L_{v_{d-1}}), \mathsf{auth}_1, \ldots, \mathsf{auth}_d$$

on a message m is verified in the natural way, by setting $L_{v_d} := m$, and then verifying $\mathsf{auth}_1, \ldots, \mathsf{auth}_d$ as in Equation (4.3).

It would then suffice for the set Y to include, say, only 160 elements of $\mathbb{Z}_N^*$ rather than 2048 elements of $\mathbb{Z}_N^*$.

Theorem 4.1. *If the RSA problem is hard relative to* GenModulus *and* GenPrime, *the Dwork-Naor scheme is existentially unforgeable under an adaptive chosen-message attack.*

Proof. Given a PPT adversary A attacking the scheme, we construct a PPT algorithm A' which attempts to solve the RSA problem. Before giving an informal overview of A', we first introduce some terminology. Say a label L_v associated with a particular node v in the tree is *legitimate* if L_v is the label assigned to v by the legitimate signer, and say the label is *illegitimate* otherwise. (We assume without loss of generality that A requests ℓ^d signatures before it outputs its forgery, and so the signer does indeed assign a label to every node in the tree.) If A outputs a valid signature forgery

$$(i_1,\ldots,i_d),(\tilde{L}_{v_1},\ldots,\tilde{L}_{v_{d-1}}),\mathrm{auth}_1,\ldots,\mathrm{auth}_d$$

on some new message m then, letting $\tilde{L}_{v_d} := m$ and $\tilde{L}_{v_0} := L_{v_0}$, there must be some minimum $r \in \{1,\ldots,d\}$ for which label $\tilde{L}_{v_{r-1}}$ is legitimate (for node v_{r-1}) but label $\tilde{L}_{v_r}$ is illegitimate (for node v_r). We will refer to v_{r-1} as the *critical node* and call i_r (indicating the position of v_r among the children of v_{r-1}) the *critical index*. If we let L_{v_r} denote the legitimate label of v_r, then there must be a minimum position $j \in \{1,\ldots,2k\}$ for which the jth bit of $\tilde{L}_{v_r}$ and the jth bit of L_{v_r} differ; call such j the *critical position*.

We are now ready to give an overview of A'. Algorithm A' is given an instance (N,y,e) of the RSA problem, and will use A as a subroutine in an attempt to solve the given instance. At the beginning of its execution, A' guesses values $i^* \in \{1,\ldots,\ell\}$ and $j^* \in \{1,\ldots,2k\}$ for the critical index and critical position, respectively; it then generates a public key and simulates the actions of a legitimate signer for A. If A outputs a forgery and the guesses i^*, j^* made by A' are correct, then A' will be able to solve its given instance of the RSA problem; since this happens with inverse polynomial probability, we see that that the success probability of A is polynomially related to the success probability of A'. We remark that A' does *not* need to guess the critical node.

We now describe A' in more detail. A' is given N,y, and a prime e, and is supposed to compute $y^{1/e} \bmod N$. To do this, it first generates a public key and node labels in the following way:

- Choose random $i^* \in \{1,\ldots,\ell\}$ and $j^* \in \{1,\ldots,2k\}$. (As described above, i^* represents a guess as to the critical index of the forgery output by A, and j^* represents a guess as to the critical position.)
- Set $e_{i^*} := e$. For $i \neq i^*$, compute $e_i \leftarrow \mathsf{GenPrime}(1^k)$; set $E := \{e_1,\ldots,e_\ell\}$. Let $\hat{e} \stackrel{\text{def}}{=} \prod_i e_i$. (We assume throughout the following that the $\{e_i\}$ are distinct, since this occurs with all but negligible probability.)
- Set $y_{j^*} := y^{\hat{e}/e_{i^*}} \bmod N$. For $j \neq j^*$, choose $s_j \leftarrow \mathbb{Z}_N^*$ and set $y_j := s_j^{\hat{e}} \bmod N$. Let $Y := \{y_1,\ldots,y_{2k}\}$.
- For each non-leaf node v in the tree, A' now generates a label L_v. It does so in a way that ensures it can compute appropriate authentication values for all nodes (except possibly the leaves), as we will discuss. For each non-leaf node v, we will let b_v denote the j^*th bit of L_v
  A' begins by choosing a random bit b_v for every leaf node v. Starting from the nodes at level $d-1$ and working backwards to the root (at level 0), A' assigns label L_v to node v as follows:

  - Let v' be the i^*th child of v, with associated bit $b_{v'}$.
  - Choose random $r_v \in \mathbb{Z}_N^*$ and compute

$$L_v := r_v^{\hat{e}}/y_{j^*}^{b_{v'}} \bmod N.$$

The bit b_v associated with node v is defined to be the j^*th bit of L_v.

- The public key is (N, L_{v_0}, Y, E), where L_{v_0} is the value assigned to the root v_0 in the above procedure.

We claim that both the public key and the labels generated in the above process are distributed exactly the same as they would be in a real execution of the signature scheme. It is immediate that N and the $\{e_i\}$ have the correct distribution. (Here, we use the fact that the prime e given as input to A' is generated using GenPrime.) Furthermore, since the initial input y as well as the $\{s_j\}$ are all uniformly distributed in $\mathbb{Z}_N^*$, and the $\{e_i\}$ are all relatively prime to $\phi(N)$, it follows that the $\{y_j\}$ are uniformly distributed as required. One can also easily check that the label of every non-leaf node is a uniform element of $\mathbb{Z}_N^*$.

We may also observe that, due to the way the $\{y_i\}$ are computed, A' can compute $y_j^{1/e_i} \bmod N$ for all $(j, i) \neq (j^*, i^*)$. In particular, then, A' is able to compute the necessary authentication value for any internal node v' (with label $L_{v'}$) having parent v (with label L_v): Indeed, if v' is the i^*th child of v then

$$\mathsf{auth}(L_v \overset{i^*}{\to} L_{v'}) \overset{\text{def}}{=} \left(L_v \cdot \prod_{j : [L_{v'}]_j = 1} y_j \right)^{1/e_{i^*}} \bmod N$$

$$= \left(r_v^{\hat{e}} \cdot \prod_{\substack{j : [L_{v'}]_j = 1 \\ j \neq j^*}} y_j \right)^{1/e_{i^*}} \bmod N,$$

using the fact that $[L_{v'}]_{j^*} \overset{\text{def}}{=} b_{v'}$. Since, by construction, A' knows the e_{i^*}th root of y_j when $j \neq j^*$, the desired authentication value can be computed. Similarly, if v' is the ith child of v for some $i \neq i^*$, then

$$\mathsf{auth}(L_v \overset{i}{\to} L_{v'}) = \left((r_v^{\hat{e}} / y_{j^*}^b) \cdot \prod_{j : [L_{v'}]_j = 1} y_j \right)^{1/e_i} \bmod N$$

for some bit b; since A' knows the e_ith root of every y_j (including y_{j^*}), the necessary authentication value can again be computed.

Furthermore, we show how A' can compute the desired answer $y^{1/e}$ if its guesses for i^* and j^* are correct. Say A outputs a forged signature which contains an illegitimate label $\tilde{L}_{v'}$ associated with a node v', where v' is the i^*th child of its parent v having legitimate label $\tilde{L}_v = L_v$. Let $L_{v'}$ denote the legitimate label of node v', and assume further that $\tilde{L}_{v'}$ and $L_{v'}$ differ on their j^*th bit. As part of its forgery, A must have output a value $\mathsf{auth} = \mathsf{auth}(L_v \overset{i^*}{\to} \tilde{L}_{v'})$ satisfying:

$$\mathsf{auth}^{e_{i^*}} = L_v \cdot \prod_{j : [\tilde{L}_{v'}]_j = 1} y_j \bmod N$$

$$= (r_v^{\hat{e}}/y_{j^*}^{b_{v'}}) \cdot \prod_{j:[\tilde{L}_{v'}]_j=1} y_j \bmod N$$

$$= r_v^{\hat{e}} \cdot y_{j^*}^{1-2b_{v'}} \cdot \prod_{\substack{j:[\tilde{L}_{v'}]_j=1 \\ j \neq j^*}} y_j \bmod N,$$

relying here on the fact that $[\tilde{L}_{v'}]_{j^*} = 1 - b_{v'}$ (since $[L_{v'}]_{j^*} = b_{v'}$, and $L_{v'}$ and $\tilde{L}_{v'}$ differ on their j^*th bit). Let $S \stackrel{\text{def}}{=} \{j : [\tilde{L}_{v'}]_j = 1 \wedge j \neq j^*\}$ and $b \stackrel{\text{def}}{=} b_{v'}$. Re-arranging, using the fact that $e = e_{i^*}$, and substituting the chosen values for the $\{y_j\}$, we obtain:

$$\left(\frac{\text{auth}}{r_v^{\hat{e}/e} \prod_{j \in S} s_j^{\hat{e}/e}} \right)^e = y^{(1-2b)\cdot(\hat{e}/e)} \bmod N.$$

Since $(1 - 2b) \in \{-1, 1\}$ and $\hat{e}/e$ is relatively prime to e, Lemma 4.1 shows that A' can efficiently compute the eth root of y, as desired.

One piece is missing in our informal description of A'. We showed earlier that A' can compute the necessary authentication values for any *internal* node; however, this does not extend to the case of computing the necessary authentication values for the *leaves*. The problem is that the labels for the leaves are outside the control of A', since A chooses "labels" for the leaves by selecting messages to be signed. We remark that it would be easy to show that the Dwork-Naor scheme is existentially unforgeable under a *known* message attack (and then apply Construction 1.2 to obtain security under a chosen-message attack); as we will see, however, the Dwork-Naor scheme can be proven secure as-is under a chosen-message attack.

To be precise about where the difficulty lies, note that for any leaf v that is *not* an i^*th child, A' can compute the desired authentication value regardless of the message m chosen by A to be assigned to this leaf. On the other hand, when a leaf v *is* an i^*th child then A' can compute the desired authentication value *only* when b_v is equal to the j^*th bit of the message m chosen by A to be assigned to this leaf. Since the view of A is independent of b_v, it will be possible for A' to generate a valid signature with probability $1/2$. Unfortunately, this is not enough to provide a "good" simulation since there are polynomially many leaves of the tree that are i^*th children (and A' would have to answer correctly for all such leaves).

Instead, we will have A' *rewind* A (using a new guess for b_v each time) in order to enable a good simulation. Say a leaf v is "hard" if it is an i^*th child of its parent, and call it "easy" otherwise. For messages associated with "easy" leaves, A' will provide a signature as discussed earlier. For a message m associated with a "hard" leaf v' whose parent is v, do:

1. If $m_{j^*} = b_{v'}$, provide a signature using the approach described earlier.
2. Otherwise, rewind A to the point where node v is first used. Choose a new, random bit $b_{v'}$ for the leaf and generate a new label L_v for the parent node v as follows:

Repeat the following until the j^*th bit of L_v is equal to b_v: choose random $r_v \in \mathbb{Z}_N^*$ and compute

$$L_v := r_v^{\hat{e}}/y_{j^*}^{b_v} \bmod N.$$

3. Resume execution of A from the point where v is first used, and return to Step 1 when signing a message associated with the leaf v'.

Note that the value of b_v is *not* changed, so no other labels are affected by the above and A' can continue to provide valid signatures until possibly reaching leaf v' again.

Assuming that A' can complete the simulation described above, the view of A at the end of the simulation is identically distributed to the view of A in an execution with a real signer. Moreover, this simulation is independent of i^* and j^*; thus, the guesses of i^*, j^* are correct with probability $1/2\ell k$. A consequence is that if A outputs a forgery with some probability ε, then A' outputs the desired RSA inverse with probability $\varepsilon' = \varepsilon/\ell k$. We thus conclude that ε is negligible.

It remains only to argue that A' can complete its simulation in (expected) polynomial time. View the simulation provided by A' as occurring in a sequence of *phases*, where a phase is identified with a node in level $d-1$ that is being used to issue signatures. (Thus, a given phase associated with a node v at level $d-1$ encompasses ℓ signatures associated with each of the ℓ children of v.) In the phase corresponding to node v (at level $d-1$), A' can issue all signatures until (possibly) the point when it reaches the i^*th child of v; at that point, A' can compute a valid signature with probability $1/2$ and must otherwise rewind A to the beginning of that phase. It is thus clear that A' can complete the simulation of any given phase in expected polynomial time. Furthermore, once A' successfully completes the simulation of some phase it proceeds to the next phase (and never rewinds to a point prior to the current phase). Since there are a polynomial number of phases, it follows that A' can complete its entire simulation in expected polynomial time.[6] This completes the proof.

4.2.2 The Cramer-Damgård Scheme

A drawback of the Dwork-Naor scheme is that the public key is relatively large, as it contains both a set E of ℓ prime numbers as well as a set Y of $2k$ elements of $\mathbb{Z}_N^*$. Cramer and Damgård [34] introduced a modification of the Dwork-Naor scheme which improves the length of the public key by avoiding the need for the set Y. (Note that ℓ is likely to be an order of magnitude smaller than k, and so this does indeed yield a significant improvement in practice.) In addition, the Cramer-Damgård scheme has some conceptual advantages as compared to the Dwork-Naor

---

[6] We can convert A' to a *strict* polynomial-time algorithm using standard techniques. Let k^c be an upper bound on the expected running time of A', and say A' succeeds with probability at least $1/k^{c'}$ for infinitely many values of k. Using Markov's inequality, A' runs more than $2k^{c'}k^c$ steps with probability less than $1/2k^{c'}$; truncating A''s execution at $2k^{c'+c}$ steps gives a strict polynomial-time algorithm with success probability at least $1/k^{c'} - 1/2k^{c'} = 1/2k^{c'}$ for infinitely many values of k.

scheme: (1) it avoids the need for rewinding in the proof of security, thus simplifying the proof; also, (2) the Cramer-Damgård scheme serves as a sort of "template" for the Cramer-Shoup signature scheme that we will see later.

The Cramer-Damgård scheme has the same underlying structure as the Dwork-Naor scheme in that it also relies on a "short" tree of "high" degree ℓ in which a node v' can be authenticated by its parent v *independently* of the other children of v. The scheme also relies on a set $E := \{e_1, \ldots, e_\ell\}$ of ℓ distinct primes to perform this authentication. (Each such prime will now be of length $2k + 1$, but this technical detail can be ignored for now.) The primary difference between the schemes — and what leads to the efficiency improvement — is the use of a different authentication technique which now requires only a single element $h \in \mathbb{Z}_N^*$ to be included in the public key (rather than a set Y of $2k$ such elements). This basic authentication step, which enables a vertex with label L to authenticate its ith child $(1 \leq i \leq \ell)$ having label L', is defined as follows:

$$\mathsf{auth}(L \xrightarrow{i} L') \stackrel{\text{def}}{=} \left(L \cdot h^{L'}\right)^{1/e_i} \bmod N, \tag{4.4}$$

where L' is a $2k$-bit string viewed as an integer in the range $\{0, \ldots, 2^{2k} - 1\}$ (in fact, all internal labels will be elements of $\mathbb{Z}_N^*$). As usual, this authentication information can be computed if the factorization of N is known. Furthermore, given a (presumably authenticated) parent vertex with label L, the candidate label L' of its ith child, and authentication information auth, anyone can verify the authenticity of L' by checking whether

$$\mathsf{auth}^{e_i} \stackrel{?}{=} L \cdot h^{L'} \bmod N. \tag{4.5}$$

Plugging the basic authentication step of Equation (4.4) into the Dwork-Naor construction yields a signature scheme which can be shown[7] to be existentially unforgeable under a *known* message attack; applying Construction 1.2 (using an arbitrary one-time signature scheme secure under a known-message attack) would then give a scheme that is existentially unforgeable under an *adaptive* chosen-message attack. Better efficiency can be obtained, however, by relying on the paradigm of Construction 1.2 but using a *specific* one-time signature scheme at the bottom level, in particular, by using at the bottom level the same basic authentication step as above (with the same values of h, N) but with a *different* prime $e_{\ell+1} \notin \{e_1, \ldots, e_\ell\}$. As in the Dwork-Naor scheme, then, the signer implicitly works with a tree of depth d having out-degree ℓ and whose root v_0 is labeled with a value L_{v_0} included in the public key. Here, however, a leaf of the tree is not directly labeled with a message to be signed, but is instead assigned a randomly generated label that is then used to authenticate the message itself. One can picture this as a tree in which there is a designated "message node" hanging off each "leaf". (For consistency with the previous

---

[7] We remark that the rewinding technique used in the proof of Theorem 4.1 would not apply in this case since one would have the guess the entire message whose signature is requested by the adversary, rather than just a single bit of this message.

notation, when a leaf with label L is used to authenticate a message m, the resulting authentication information will be denoted by $\mathsf{auth}(L \overset{\ell+1}{\rightarrow} m)$.)

4.2.2.1 A One-Time Signature Scheme

Before giving a complete description and proof of security for the full Cramer-Damgård scheme, it will be instructive to analyze the basic authentication step as a one-time signature scheme. (See Construction 4.2.) We do not directly rely on this result in what follows, but the techniques used here will be helpful in understanding the proof of the full Cramer-Damgård scheme.

Construction 4.2: An RSA-based one-time signature scheme

Let $\mathsf{GenModulus}, \mathsf{GenPrime}$ be as described in the text.

Key generation: On security parameter 1^k, proceed as follows:

- Compute $(N, p, q) \leftarrow \mathsf{GenModulus}(1^k)$ and $e \leftarrow \mathsf{GenPrime}(1^{2k+1})$.
- Choose random $L_{v_0}, h \in \mathbb{Z}_N^*$.
- The public key is (N, L_{v_0}, h, e) and the secret key is d.

Signature generation: To sign message $m \in \{0,1\}^{2k}$, viewed as an integer in the range $\{0, \ldots, 2^{2k} - 1\}$, the signer computes

$$\mathsf{auth} := \left(L_{v_0} \cdot h^m \right)^{1/e}.$$

The signature is auth.

Signature verification: To verify signature auth on message m, simply verify whether $\mathsf{auth}^e \overset{?}{=} L_{v_0} \cdot h^m \bmod N$.

Theorem 4.2. *If the RSA problem is hard relative to* $\mathsf{GenModulus}$ *and* $\mathsf{GenPrime}$, *then Construction 4.2 is strongly unforgeable under a one-time, known-message attack.*

Proof. Since for a given public key there is only one valid signature on any given message, it suffices to prove (regular) unforgeability. Given a PPT adversary A which forges a valid signature on a new message with non-negligible probability, we construct a PPT algorithm A' that solves the RSA problem with the same probability. A', given an RSA modulus N, an element $y \in \mathbb{Z}_N^*$, and a $(2k+1)$-bit prime e, uses A as a subroutine in the following way: First, run $A(1^k)$ to obtain a message m to be signed. Set $h := y$, choose random $s \in \mathbb{Z}_N^*$, and compute $L_{v_0} := s^e h^{-m} \bmod N$. Give to A the public key (N, L_{v_0}, h, e) along with the signature s. One can verify that the public key given to A is distributed identically to a real public key (in particular, L_{v_0} is uniformly distributed in $\mathbb{Z}_N^*$), and s is a valid signature on m with respect to the given public key.

We claim that if A outputs (m', s') such that $m' \neq m$ and $(s')^e = L_{v_0} h^{m'} \bmod N$ (i.e., A forges a valid signature s' on a new message m'), then A' can compute $y^{1/e} \bmod N$. To see this, assume without loss of generality that $m' > m$ and observe that

$$(s')^e h^{-m'} = L_{v_0} = s^e h^{-m} \bmod N$$

and so $(s'/s)^e = h^{m'-m} = y^{m'-m} \bmod N$. Since $m, m' \in \{0, \ldots, 2^{2k} - 1\}$, their difference $m' - m$ is less then e (here, we use the fact that e is a $(2k+1)$-bit integer). Since e is prime, it follows that $\gcd(e, m' - m) = 1$ and so Lemma 4.1 applies. But this means A' can efficiently compute the eth root of y modulo N.

4.2.2.2 The Cramer-Damgård Scheme

The Cramer-Damgård scheme is given as Construction 4.3. As in the case of the Dwork-Naor scheme, the scheme as described is stateful (but see footnote 5).

Theorem 4.3. *If the RSA problem is hard relative to* GenModulus *and* GenPrime, *then the Cramer-Damgård scheme is existentially unforgeable under an adaptive chosen-message attack.*

Proof. We first establish some notation. The nodes at the first $d - 1$ levels of the tree will be referred to as *internal nodes*, while the nodes at depth d will be called *leaves*. The union of these will be called the *tree nodes*. The children of the leaves will be referred to as *message nodes*. As in the proof of Theorem 4.1, a label L_v associated with particular node v is called *legitimate* if L_v is the label assigned to this node by the signer; the label is *illegitimate* otherwise. (As before, we assume without loss of generality that the adversary requests ℓ^d signatures before it outputs its forgery, and so the signer does indeed assign a label to every tree node.)

Let A be a PPT adversary attacking the Cramer-Damgård signature scheme and having success probability $\varepsilon(k)$. If A outputs a valid forgery

$$(i_1, \ldots, i_d), (\tilde{L}_{v_1}, \ldots, \tilde{L}_{v_d}), \mathsf{auth}_1, \ldots, \mathsf{auth}_{d+1} \tag{4.6}$$

on some message m then, letting $\tilde{L}_{v_{d+1}} := m$ and $\tilde{L}_{v_0} := L_{v_0}$, there must be some minimum $r \in \{1, \ldots, d+1\}$ for which label $\tilde{L}_{v_{r-1}}$ is legitimate (for node v_{r-1}) but $\tilde{L}_{v_r}$ is illegitimate (for node v_r). We will refer to such v_{r-1} as the *critical node*.

Let $\varepsilon_1(k)$ denote the probability that A outputs a signature forgery whose critical node is an internal node, and let $\varepsilon_2(k)$ be the probability that A outputs a signature forgery whose critical node is a leaf. We will give PPT algorithms A_1', A_2' that run A as a subroutine and solve the RSA problem with probability polynomially related to $\varepsilon_1, \varepsilon_2$, respectively. Because the RSA problem is hard for GenRSA, both ε_1 and ε_2 must therefore be negligible. Since $\varepsilon(k) = \varepsilon_1(k) + \varepsilon_2(k)$, this yields the theorem.

Construction 4.3: The Cramer-Damgård scheme

Let GenModulus, GenPrime be as described in the text.

Key generation: On security parameter 1^k, proceed as follows:

- Run $(N, p, q) \leftarrow$ GenModulus(1^k).
- For $i = 1$ to $\ell + 1$, compute $e_i \leftarrow$ GenPrime(1^{2k+1}). Set $E := \{e_1, \ldots, e_{\ell+1}\}$.
- Choose random $L_{v_0}, h \in \mathbb{Z}_N^*$.
- The public key is (N, L_{v_0}, h, E) and the secret key is p, q.

Signature generation: The signer implicitly maintains a tree of depth d and out-degree ℓ whose root v_0 is labeled with L_{v_0}. The ℓ^d "leaves" of this tree each have a single child at depth $d + 1$ (and thus, technically speaking, they are not leaves although we will continue to refer to them as such). Each leaf will be used to sign a single message.

To generate a signature on a message $m \in \{0, 1\}^{2k}$, let v_d be the left-most leaf in the tree which has not yet been used, and let v_{d+1} be its child. Assign a random label $L_{v_d} \in \mathbb{Z}_N^*$ to v_d, and set $L_{v_{d+1}} := m$. Let $(i_1, \ldots, i_d)$ denote the sequence of edges on the path from the root to the leaf (where each $i_1, \ldots, i_d$ lies in the range $\{1, \ldots, \ell\}$), and let $(v_1, \ldots, v_d)$ be the nodes on this path (not including the root). Let $(L_{v_1}, \ldots, L_{v_d})$ denote the labels of these nodes; if any of these nodes have not yet been assigned a label, they are now assigned a label chosen uniformly at random from $\mathbb{Z}_N^*$. The signature is then:

$$(i_1, \ldots, i_d), (L_{v_1}, \ldots, L_{v_d}),$$
$$\mathsf{auth}(L_{v_0} \overset{i_1}{\to} L_{v_1}), \ldots, \mathsf{auth}(L_{v_{d-1}} \overset{i_d}{\to} L_{v_d}), \mathsf{auth}(L_{v_d} \overset{\ell+1}{\to} L_{v_{d+1}}),$$

where the $\mathsf{auth}(L \overset{i}{\to} L')$ are computed as in Equation (4.4).

Signature verification: A signature

$$(i_1, \ldots, i_d), (L_{v_1}, \ldots, L_{v_d}), \mathsf{auth}_1, \ldots, \mathsf{auth}_d, \mathsf{auth}_{d+1}$$

on a message m is verified in the natural way, by using the root value L_{v_0} contained in the public key, setting $L_{v_{d+1}} := m$, and then verifying the values $\mathsf{auth}_1, \ldots, \mathsf{auth}_{d+1}$ as in Equation (4.5).

Algorithm A_1' is constructed using essentially the same idea used to construct algorithm A' in the proof of the Dwork-Naor scheme; actually, the proof is a bit simpler here since *all legitimate labels of the tree nodes are chosen uniformly at random by the signer* (and, in particular, are outside the control of A); thus, no rewinding of A is necessary. Algorithm A_2' is devised in a manner similar to the A' used in the proof of the one-time signature scheme of the previous section (cf. Theorem 4.2) using the observation that, in the present scheme, the signer does not "commit" to the label L_v of any leaf v until *after* the adversary decides what message should be authenticated by this leaf. (This is in contrast to the Dwork-Naor scheme where the label of a node v at level $d-1$ is chosen and revealed by the signer *before* all the messages that will be authenticated by the children of v are determined. Note the similarity with Construction 1.2.)

We proceed with a description and analysis of the two algorithms. In each case we use the adversary A as a subroutine of an algorithm that will attempt to compute $y^{1/e} \bmod N$ for given input values N, y, and e.

Algorithm A_1'. Say A outputs a forgery as in Equation (4.6). If the critical node v_{r-1} is an internal node then we call i_r the *critical index*. As in the proof of Theorem 4.1, we will have A_1' guess in advance the value $i^* \in \{1, \dots, \ell\}$ of the critical index. A_1' will then generate a public key (N, L_{v_0}, h, E) along with labels for all the tree nodes such that (1) these values are distributed identically to the public key and tree-node labels in a real execution of the signature scheme; (2) A_1' will be able to answer all signing queries of A (without having to rewind A); yet (3) if A outputs a forgery whose critical node is an internal node, and A_1''s guess of the critical index is correct, then A_1' will be able to compute the desired answer $y^{1/e} \bmod N$. Since the guess of the critical index is correct with probability $1/\ell$, we conclude that A_1' succeeds in solving its given RSA instance with probability $\varepsilon_1(k)/\ell$.

We now describe A_1' in more detail. A_1' is given N, y, and a $(2k+1)$-bit prime e, and its goal is to compute $y^{1/e} \bmod N$. It prepares a public key and labels for the tree nodes as follows:

- Choose random $i^* \leftarrow \{1, \dots, \ell\}$. Set $e_{i^*} := e$, and compute $e_i \leftarrow \mathsf{GenPrime}(1^{2k+1})$ for $i \neq i^*$. Define $E = \{e_1, \dots, e_\ell, e_{\ell+1}\}$, and let $\hat{e} := \prod_{i=1}^{\ell+1} e_i$. (We assume in what follows that all the primes $\{e_i\}$ are distinct, since this occurs with all but negligible probability.)
- Set $h := y^{\hat{e}/e_{i^*}} \bmod N$.
- Generate labels for all tree nodes in the following bottom-up fashion:
  - For a node v at level d (i.e., v is a leaf node), choose the label L_v as follows: pick random $s_v \leftarrow \mathbb{Z}_N^*$ and let $L_v := s_v^{e_{\ell+1}} \bmod N$.
  - For a node v at level $r < d$, choose the label L_v as follows: let w be the i^*th child of v having label L_w. Choose $s_v \leftarrow \mathbb{Z}_N^*$ and set $L_v := s_v^{\hat{e}} \cdot h^{-L_w} \bmod N$.

  In this way, A_1' eventually obtains a label L_{v_0} for the root of the tree.
- The public key is (N, L_{v_0}, h, E).

We first claim that both the public key and the labels generated in the above process are distributed identically to their distribution in a "real" execution of the signature scheme. This is immediate for the case of N and the $\{e_i\}$. Since y is a uniform element of $\mathbb{Z}_N^*$ and $\hat{e}/e_{i^*}$ is relatively prime to $\phi(N)$, we see that h is uniformly distributed as well. A similar argument applies for the labels of all the tree nodes since the $\{s_v\}$ are chosen independently and uniformly at random from $\mathbb{Z}_N^*$.

More interestingly, the public key and the labels of the tree nodes are set up in such a way that A_1' can compute a valid signature on any message given to it by the adversary (without rewinding); this follows from the observations that:

- Say $i \neq i^*$, and let w be the ith child of some internal node v. Then A_1' can compute $\mathsf{auth}(L_v \overset{i}{\to} L_w)$, where L_v, L_w are the labels computed in the above process. To see this, let w^* be the i^*th child of v and note that

$$\mathsf{auth}(L_v \overset{i}{\to} L_w) \overset{\text{def}}{=} \left(L_v \cdot h^{L_w}\right)^{1/e_i} \bmod N$$
$$= \left(s_v^{\hat{e}} \cdot h^{L_w - L_{w^*}}\right)^{1/e_i} \bmod N$$
$$= s_v^{\hat{e}/e_i} \cdot \left(y^{\hat{e}/e_i e_{i^*}}\right)^{L_w - L_{w^*}} \bmod N.$$

Since both $\hat{e}/e_i$ and $\hat{e}/e_i e_{i^*}$ are integers, A_1' can compute the above even though it does not know the factorization of N.

- Let w^* be the i^*th child of some internal node v. We show that A_1' can compute $\mathsf{auth}(L_v \overset{i^*}{\to} L_{w^*})$, where L_v, L_{w^*} are the labels computed in the above process. To see this, note that:

$$\mathsf{auth}(L_v \overset{i^*}{\to} L_{w^*}) \overset{\text{def}}{=} \left(L_v \cdot h^{L_{w^*}}\right)^{1/e_{i^*}} \bmod N$$
$$= \left(s_v^{\hat{e}}\right)^{1/e_{i^*}} \bmod N$$
$$= s_v^{\hat{e}/e_{i^*}} \bmod N.$$

Again, since $\hat{e}/e_{i^*}$ is an integer, A_1' can compute the above without knowledge of the factorization of N.

- For any leaf v and any message m chosen by the adversary to be authenticated by this leaf, A_1' can compute $\mathsf{auth}(L_v \overset{\ell+1}{\to} m)$. To see this, note that

$$\mathsf{auth}(L_v \overset{\ell+1}{\to} m) = (L_v \cdot h^m)^{1/e_{\ell+1}} \bmod N$$
$$= \left(s_v^{e_{\ell+1}} \cdot \left(y^{\hat{e}/e_{i^*}}\right)^m\right)^{1/e_{\ell+1}} \bmod N$$
$$= s_v \cdot \left(y^{\hat{e}/e_{\ell+1} e_{i^*}}\right)^m \bmod N.$$

Once again, since $\hat{e}/e_{\ell+1} e_{i^*}$ is an integer, A_1' can easily compute the above.

Finally, we show that if A outputs a valid forgery where the critical node is an internal node, and the guess i^* for the critical index is correct, then A_1' can compute the desired solution $y^{1/e} = y^{1/e_{i^*}} \bmod N$. Indeed, assume A outputs a valid forgery as in Equation (4.6) with critical node v_{r-1} (for $r \leq d$) and such that $i_r = i^*$. Using the fact that the forgery is valid we have (recall that L_{v_r} is the legitimate label of node v_r):

$$\mathsf{auth}_r^{e_{i^*}} = L_{v_{r-1}} \cdot h^{\tilde{L}_{v_r}} \bmod N$$
$$= \left(s_{v_{r-1}}^{\hat{e}} \cdot h^{-L_{v_r}}\right) \cdot h^{\tilde{L}_{v_r}} \bmod N$$
$$= s_{v_{r-1}}^{\hat{e}} \cdot h^{\tilde{L}_{v_r} - L_{v_r}} \bmod N$$

$$= s_{v_{r-1}}^{\hat{e}} \cdot y^{(\hat{e}/e_{i^*}) \cdot (\tilde{L}_{v_r} - L_{v_r})} \bmod N,$$

and so A_1' can compute the value $\gamma := \text{auth}_r / s_v^{\hat{e}/e_{i^*}} \bmod N$ such that

$$\gamma^{e_{i^*}} = y^{(\hat{e}/e_{i^*}) \cdot (\tilde{L}_{v_r} - L_{v_r})} \bmod N.$$

Now, since e_{i^*} is relatively prime to both $\hat{e}/e_{i^*}$ and $|\tilde{L}_{v_r} - L_{v_r}|$ (using, in the latter case, the fact that $|\tilde{L}_{v_r} - L_{v_r}| < e_{i^*}$ since e_{i^*} is a $(2k+1)$-bit prime while $\tilde{L}_{v_r}$ and L_{v_r} are $2k$-bit values), A_1' can compute $y^{1/e_{i^*}}$ using Lemma 4.1.

It is easy to see that A's view is independent of the initial guess i^* made by A_1'. Thus, the probability that this guess is correct, even conditioned on the fact that A outputs a forgery whose critical node is an internal node, is exactly $1/\ell$. This means that A_1' solves its given RSA instance with probability $\varepsilon_1(k)/\ell$. Since this must be negligible by assumption, we conclude that ε_1 is negligible.

Algorithm A_2'. We now construct an algorithm A_2' that succeeds in solving its given RSA instance whenever A outputs a valid forgery whose critical node is a leaf. Now, A_2' will construct a public key along with a set of node labels for all internal nodes (but not the leaves) in such a way that A_2' can authenticate any leaf node *regardless of the label this leaf nodes is assigned*; to authenticate a message chosen by A, we will then have A_2' choose the label of the corresponding leaf in such a way that it can issue a legitimate signature. It is thus essential that in the Cramer-Damgård scheme the label of the leaf used to authenticate some message is not determined until *after* the message to be signed is chosen by A. This is in contrast to the Dwork-Naor scheme (where the adversary gets to see the label of a node before having to decide on some message that should be authenticated by that node), and is essential to the current proof.

We now give a more complete description of A_2'. Recall that A_2' is given N, y, e and is to compute $y^{1/e} \bmod N$. It prepares a public key and labels for the internal nodes of the tree as follows:

- Set $e_{\ell+1} := e$ and compute $e_i \leftarrow \text{GenPrime}(1^{2k+1})$ for $i \neq i^*$. Define $E := \{e_1, \dots, e_\ell, e_{\ell+1}\}$, and let $\hat{e} := \prod_{i=1}^{\ell} e_i$ (note that $\hat{e}$ is defined differently than before). We assume in what follows that all the primes $\{e_i\}$ are distinct, as this occurs with all but negligible probability.
- Set $h := y^{\hat{e}} \bmod N$.
- Generate a label L_v for every internal node v as follows: pick random $s_v \leftarrow \mathbb{Z}_N^*$ and let $L_v := s_v^{\hat{e}} \bmod N$.
- The public key is (N, L_{v_0}, h, E), where v_0 is the root of the tree.

Using similar arguments as in the previous analysis of A_1', one can verify that the public key and all node labels generated in the above procedure are distributed exactly as in a real execution of the signature scheme. We thus turn to showing that A_2' can indeed authenticate any leaf regardless of the label that leaf is assigned. Indeed, let v be an internal node and let w be the ith child of v for some $i \in \{1, \dots, \ell\}$ (note that w may be a leaf). Then regardless of the label L_w assigned to w, the simulator can compute $\text{auth}(L_v \xrightarrow{i} L_w)$ since:

$$\mathsf{auth}(L_v \overset{i}{\to} L_w) \overset{\text{def}}{=} \left(L_v \cdot h^{L_w}\right)^{1/e_i} \bmod N$$

$$= \left(s_v^{\hat{e}} \cdot y^{\hat{e}}\right)^{1/e_i} \bmod N$$

$$= s_v^{\hat{e}/e_i} \cdot y^{\hat{e}/e_i} \bmod N,$$

and $\hat{e}/e_i$ is an integer.

Given the above, it is straightforward to show that A_2' can respond correctly to any signing query of A. This will be done as in the proof of Theorem 4.2: when A requests (by making a signing query) that a message m be authenticated by leaf v, algorithm A_2' chooses $s_v \leftarrow \mathbb{Z}_N^*$ and sets the label of v to

$$L_v := s_v^{e_{\ell+1}} \cdot h^{-m} \bmod N.$$

Note that L_v is distributed uniformly at random in $\mathbb{Z}_N^*$, as in a real execution of the signature scheme. Now, by what we have said earlier, A_2' can provide correct authentication values for each of the nodes on the path from the root to v; it only remains to show that A_2' can authenticate m using the newly created label L_v. This is done by setting $\mathsf{auth}(L_v \overset{\ell+1}{\to} m) := s_v$ (it is easily verified that this is the correct authentication value).

Say A outputs a forgery in which the critical node is a leaf v. To complete the description of A_2', we show how this allows A_2' to compute the desired solution $y^{1/e} \bmod N$. Let m be the message that A_2' authenticated using v, let m' be the message included by A in its forgery, and let L_v denote the legitimate label of v (constructed as described above). As in the proof of Theorem 4.2, A_2' can recover from A's forgery the value

$$s \overset{\text{def}}{=} \mathsf{auth}(L_v \overset{\ell+1}{\to} m')$$

$$= \left(L_v \cdot h^{m'}\right)^{1/e_{\ell+1}} \bmod N$$

$$= \left(s_v^{e_{\ell+1}} \cdot y^{(m'-m)\cdot\hat{e}}\right)^{1/e_{\ell+1}} \bmod N$$

$$= s_v \cdot y^{(m'-m)\cdot\hat{e}/e_{\ell+1}} \bmod N,$$

and so s/s_v is the $e_{\ell+1}$th root of $y^{(m'-m)\cdot\hat{e}}$ (recall that A_2' knows s_v). Because $m', m \in \{0, \dots, 2^{2k} - 1\}$, we have $|m - m'| < e_{\ell+1}$; since $e_{\ell+1}$ is prime, it follows that $\gcd(e_{\ell+1}, |m - m'|) = 1$. We know also that $e_{\ell+1}$ and $\hat{e}$ are relatively prime. Using Lemma 4.1, it follows that A_2' can compute the desired $e_{\ell+1}$th root of y.

To conclude, we have shown that A_2' computes the desired solution with probability exactly $\varepsilon_2(k)$, which must be negligible by assumption. It must therefore be the case that ε_2 is negligible, as desired.

As we have shown that both ε_1 and ε_2 are negligible, this completes the proof of the theorem.

4.2.3 The Hohenberger-Waters Scheme

Hohenberger and Waters [67] recently introduced a signature scheme based on the RSA assumption that has several advantages relative to the two schemes described previously. The primary advantage is that signatures in the Hohenberger-Waters scheme are *short*, even though a tree of large depth is still (implicitly) used. (Even though the Hohenberger-Waters scheme is only proved security under a known-message attack, one can apply Construction 1.2 to get a scheme secure against chosen-message attacks whose signatures are still shorter than in the Dwork-Naor or Cramer-Damgård schemes.) Because the signature length is independent of the tree depth, and the tree is maintained only implicitly, the Hohenberger-Waters scheme can afford to use a binary tree of depth k (and exponential size) and can therefore more easily be made stateless. (Specifically, a leaf can be assigned to each possible message $m \in \{0,1\}^k$ to be signed, as in Construction 3.2.)

In the Dwork-Naor and Cramer-Damgård schemes, the signer uses a tree of degree ℓ and a prime e_i is associated with each of the ℓ outgoing "directions". Thus, for example, the *same* prime e_1 is assigned to the left-most outgoing edge of every node in the tree. In the Hohenberger-Waters scheme a *different* prime is associated with every edge in the tree. Since (as noted earlier) the scheme uses a tree of exponential size, it is impossible to explicitly list all these primes as part of the public key; instead, a (compact, keyed) function f mapping edges to primes is included in the public key. (The same idea could be used to reduce the size of the public key in the Dwork-Naor and Cramer-Damgård schemes.) As we will see in the proof of security, this function can be "programmed" by a simulator so as to map a specific edge to a given prime e (that is given to the simulator as part of the RSA challenge).

We now describe the keyed function f that we use.[8] Let F be a pseudorandom function that, for simplicity, we assume maps arbitrary-length inputs to k-bit outputs. Given a string $m \in \{0,1\}^{\leq k}$, define $f_{K,c}(m) = \mathsf{GenPrime}(1^k; F_K(m) \oplus c)$. The Hohenberger-Waters scheme is given as Construction 4.4.

Theorem 4.4. *If the RSA problem is hard relative to* $\mathsf{GenModulus}$ *and* $\mathsf{GenPrime}$, *even for public coins, then the Hohenberger-Waters scheme is strongly unforgeable under a known-message attack.*

Proof. Observe that for a given public key, each message m has a unique valid signature. (This relies on the fact that $\mathsf{GenPrime}$ always outputs an e relatively prime to $\phi(N)$.) We prove existential unforgeability, and strong unforgeability follows.

Fixing K, c, define $E(m) = \{f_{K,c}(m|_i)\}_{i=1}^k$; this is just the set of primes that are used to sign m. We also define $e(m) = \prod_{e \in E(m)} e$ (this matches the definition of $e(m)$ in the description of the scheme). If we imagine that the signer implicitly maintains a binary tree of depth k, we can associate any message $m \in \{0,1\}^k$, in the natural way, with a path $P(m)$ in this tree from the root to a leaf. We can similarly associate any prefix $m|_i$ with a path from the root to a node at depth i; it is then natural to view the prime $e = f_{K,c}(m|_i)$ as being associated with the last edge on this path.

---

[8] Hohenberger-Waters suggest a different function, but our choice of f yields a simpler proof.

Construction 4.4: The Hohenberger-Waters scheme

Let GenModulus, GenPrime, and f be as described in the text. Given a binary string m, let $m|_i \overset{\text{def}}{=} m_1 \cdots m_i$ denote the i-bit prefix of m.

Key generation: On security parameter 1^k, proceed as follows:

- Compute $(N, p, q) \leftarrow \mathsf{GenModulus}(1^k)$.
- Choose random $h \in \mathbb{Z}_N^*$, and $K, c \in \{0, 1\}^k$.
- The public key is (N, h, K, c) and the secret key is p, q.

Signature generation: To sign message $m \in \{0, 1\}^k$, do:

- For $i = 1$ to k, let $e_i := f_{K,c}(m|_i)$. Define $e(m) = \prod_{i=1}^k e_i$.
- Output the signature $\sigma = h^{1/e(m)} \bmod N$.

Signature verification: To verify signature σ on message m, do:

- For $i = 1$ to k, let $e_i := f_{K,c}(m|_i)$. Define $e(m) = \prod_{i=1}^k e_i$.
- Output 1 iff $\sigma^{e(m)} \overset{?}{=} h \bmod N$.

Given a PPT algorithm A attacking the scheme in a known-message attack, we construct a PPT algorithm A' attempting to solve the RSA problem: A' is given as input N, e, y, ω with $\mathsf{GenPrime}(1^k; \omega) = e$. It proceeds as follows:

- Run $A(1^k)$ to get messages $m_1, \ldots, m_\ell$.
- Imagining a depth-k binary tree as discussed above, let $P_i = P(m_i)$ be the path associated with m_i and set $P := \cup_{i=1}^\ell P_i$.
- Choose a node v^* at random[9] from among those nodes adjacent to P (i.e., nodes connected to P but not in P), and set m^* equal to the message prefix associated with the path from the root to node v^*. Let j^* denote the depth of v^* (or, equivalently, the length of m^*).
- Choose random $K \in \{0, 1\}^k$, and set $c := F_K(m^*) \oplus \omega$. (Note that this ensures $f_{K,c}(m^*) = e$.)
- Let $\hat{E} = \cup_{i=1}^\ell E(m_i)$, and set $\hat{e} := \prod_{e_i \in \hat{E}} e_i$. Security of the pseudorandom function can be shown to imply that, with all but negligible probability, $e \notin \hat{E}$ (and so $\gcd(e, \hat{e}) = 1$); we assume this to be the case from now on.
- Set $h := y^{\hat{e}} \bmod N$ and give to A the public key (N, h, K, c) and the signatures $\{\sigma_i = y^{\hat{e}/e(m_i)} \bmod N\}_{i=1}^\ell$.
- If A outputs a forgery (m, σ), let v_m denote the first node on the path $P(m)$ that is not in P. (The path from the root to v_m corresponds to the shortest prefix of m

---

[9] Even though the (implicit) tree has exponential size, it is not hard to see that this step can be performed efficiently.

that is not a prefix of any of the $\{m_i\}$. Since $m \notin \{m_i\}$, some such prefix must exist.) If $v_m = v^*$, then A' computes the desired solution as described next.

- Assuming $v_m = v^*$ and recalling that v^* is at depth j^*, we thus have $m|_{j^*} = m^*$ and so

$$
e(m) = \prod_{i=1}^{\ell} f_{K,c}(m|_i) = f_{K,c}(m^*) \cdot \prod_{i \neq j^*} f_{K,c}(m|_i)
$$
$$
= e \cdot \prod_{i \neq j^*} f_{K,c}(m|_i).
$$

Because σ is a valid signature on m, we have $\sigma^{e(m)} = h \bmod N$ and so

$$
\left(\sigma^{\prod_{i \neq j^*} f_{K,c}(m|_i)} \right)^e = y^{\hat{e}} \bmod N.
$$

Using that fact that $\gcd(e, \hat{e}) = 1$, it follows from Lemma 4.1 that A' can efficiently compute the desired solution $y^{1/e} \bmod N$.

It is not hard to verify that the view of A when run by A' is identically distributed to the view of A in a real execution of the signature scheme. Moreover, since there are only polynomially many nodes adjacent to P, the guess of m^* by A' is correct with inverse polynomial probability. We conclude that A' correctly solves its given RSA instance with probability that is polynomially related to the success probability of A in its attack. This completes the proof of the theorem.

4.3 Schemes Based on the Strong RSA Assumption

The constructions we have seen so far in this chapter are more efficient than the generic schemes of the previous chapter, but are still not considered efficient enough to be used in practice (and they are certainly not competitive with the RSA-based solutions that will be discussed in Chapter 7). For this reason, researchers have explored a variant of the RSA assumption (termed the *strong RSA assumption*), and have used it to construct more efficient signature schemes. We will describe two such approaches here. The first approach, initiated by Cramer and Shoup with subsequent efficiency improvements by Fischlin, may be viewed as a modification of the Cramer-Damgård scheme in which the tree has depth 1 and the edges leaving the root are associated with primes $e_1, \ldots$ chosen "on-the-fly" (and, in particular, no longer included as part of the public key). The second approach, due to Gennaro, Halevi, and Rabin, uses the strong RSA assumption to construct a secure scheme "directly", in a manner inspired by Lemma 4.2. The Gennaro-Halevi-Rabin scheme has the advantage of being quite straightforward to analyze (at least for the variant presented here); the Cramer-Shoup/Fischlin schemes, however, appear to be more practical.

4.3.1 The Strong RSA Assumption

The RSA assumption states, informally, that given (y, e, N) it is "hard" to compute $y^{1/e} \bmod N$. The strong RSA assumption asserts that this problem remains "hard" even given the freedom to choose e: that is, given (N, y) it is "hard" to output a valid solution $(e, y^{1/e} \bmod N)$ for *any* chosen $e \geq 2$ (we sometimes refer to this as finding a *non-trivial root* of y). More formally, let GenModulus be as in the previous section. Then:

Definition 4.3. The **strong RSA problem is hard relative to** GenModulus if the following is negligible for all PPT algorithms A:

$$\Pr\left[\begin{array}{l} (N, p, q) \leftarrow \mathsf{GenModulus}(1^k); y \leftarrow \mathbb{Z}_N^*; \\ (x, e) \leftarrow A(N, y) \end{array} : e \geq 2 \bigwedge x^e = y \bmod N \right].$$

We stress that e need not be prime, nor do we require $\gcd(e, \phi(N)) = 1$.

Although the strong RSA assumption can be formulated with respect to arbitrary moduli N, for technical reasons we will assume throughout this section that the factors p, q of moduli N output by GenModulus are such that $(p-1)/2$ and $(q-1)/2$ are also prime (p, q of this type are known as *strong* primes). This ensures that $\phi(N) = (p-1)(q-1) = 4p'q'$ for p', q' prime. Note also that any odd e of length less than $k-1$ is relatively prime to $\phi(N)$ in this case.

We continue to let GenPrime denote an algorithm that, on input 1^ℓ, outputs an ℓ-bit prime. We assume primes output by GenPrime are "random" in the sense discussed at the beginning of Section 4.2.

4.3.2 Security Against Known-Message Attacks

As an instructive prelude to the full Cramer-Shoup scheme, we first present a scheme that is secure against known-message attacks. The scheme is essentially a variant of the one-time signature scheme shown in Section 4.2.2.1, where here a *fresh* prime is used each time a signature is generated. (There are some other, more minor differences as well).

Theorem 4.5. *If the strong RSA problem is hard relative to* GenModulus, *then Construction 4.5 is strongly unforgeable under a known-message attack.*

Proof. Let A be a PPT adversary attacking the scheme and having success probability $\varepsilon(k)$. A attacks the scheme in a known-message attack by requesting signatures on $t = t(k)$ messages $\{m_i\}_{i=1}^t$ chosen by A before it is given the public key. Let the public key be (N, L_{v_0}, h) and denote the signature on message m_i by (e_i, auth_i). We will assume throughout the rest of the proof that the $\{e_i\}$ are distinct (since this fails to hold with only negligible probability); therefore, if the adversary outputs a valid forgery (e, auth) on some message m, there are two cases: (1) $e = e_j$ for some

---

Construction 4.5: A scheme secure against known-message attacks

---

Let GenModulus, GenPrime be as described in the text, and set $\ell = \lfloor k/2 \rfloor$.

Key generation: On security parameter 1^k, proceed as follows:

- Run $(N, p, q) \leftarrow$ GenModulus(1^k), where p and q are strong primes.
- Choose random $L_{v_0}, h \in QR_N$.
- The public key is (N, L_{v_0}, h) and the secret key is p, q.

Signature generation: To sign message $m \in \{0, 1\}^\ell$, viewed as an integer in the range $\{0, \ldots, 2^\ell - 1\}$, the signer sets $e \leftarrow$ GenPrime$(1^{\ell+1})$ and computes

$$\mathsf{auth} := \left(L_{v_0} \cdot h^m \right)^{1/e} \bmod N$$

using the factorization of N. The signature is (e, auth).

Signature verification: To verify signature (e, auth) on message m, check that e is an odd, $(\ell + 1)$-bit number and then verify whether

$$\mathsf{auth}^e \stackrel{?}{=} L_{v_0} \cdot h^m \bmod N.$$

(unique) j, or (2) $e \notin \{e_i\}$. Let $\varepsilon_1(k)$ denote the probability of the first event, and $\varepsilon_2(k)$ be the probability of the second. We claim that both $\varepsilon_1, \varepsilon_2$ are negligible. Since $\varepsilon = \varepsilon_1 + \varepsilon_2$ (except for a negligible term relating to the probability that $e_i = e_j$ for some $i \neq j$), this concludes the proof of the theorem.

The proof that ε_1 is negligible follows almost immediately from the analysis of algorithm A_1' in the proof of Theorem 4.3 (and relies on the "standard" RSA assumption as used there); we therefore focus on bounding ε_2. (Interestingly, here we can bound ε_2 even for an adaptive chosen-message attack.) In doing so, we will rely on the strong RSA assumption. Thus, we will present a PPT algorithm A' (using A as a subroutine) that is given (N, y) as input and outputs $(e, y^{1/e})$ (with $e \geq 2$) with probability polynomially related to ε_2. Under the strong RSA assumption, it follows that ε_2 is negligible as claimed.

Let t be a polynomial upper-bound on the number of signatures requested by A. Algorithm A', on input (N, y), proceeds as follows:

- For $i = 1$ to t, compute $e_i \leftarrow$ GenPrime$(1^{\ell+1})$, and set $\hat{e} := \prod_{i=1}^t e_i$.
- Set $h := y^{2\hat{e}} \bmod N$. Choose random $a \leftarrow \{1, \ldots, N^2\}$ and set $L_{v_0} := h^a \bmod N$.
- Give to A the public key (N, L_{v_0}, h). When A requests a signature on the ith message m_i, compute

$$\mathsf{auth}_i := y^{2a\hat{e}/e_i} y^{2m_i\hat{e}/e_i} \bmod N$$

$$= \left(L_{v_0} h^{m_i}\right)^{1/e_i} \bmod N,$$

and give to A the signature (e_i, auth_i).

- If A outputs valid signature forgery (e, auth) on a message m, with $e \notin \{e_1, \ldots e_t\}$, then A' will attempt to compute $(e', y^{1/e'})$ for some $e' \geq 2$ as discussed below.

We will show in a moment how A' attempts to compute $(e', y^{1/e'})$ in the case when A outputs a valid signature forgery with $e \notin \{e_1, \ldots, e_t\}$. Let us first argue, however, that the probability that A outputs such a forgery is negligibly close to ε_2. This follows from the fact that A's view in the above interaction with A' is statistically close to A's view in an interaction with a real signer: N is clearly distributed identically in both cases, and since y is uniformly distributed in $\mathbb{Z}_N^*$ and $\gcd(\hat{e}, \phi(N)) = 1$, we have that h is uniformly distributed in QR_N. Furthermore, the signatures given to A are all valid and correctly distributed signatures with respect to the given public key. It only remains to argue that the distribution of L_{v_0} in the above experiment is statistically close to the distribution that results from the true key-generation process. The real key-generation algorithm chooses L_{v_0} uniformly from QR_N. Turning to the above experiment, we first note that QR_N is a cyclic subgroup of $\mathbb{Z}_N^*$ with order $p'q'$ for some primes p', q' (this is due to the fact that N is a product of strong primes). Since h is a random quadratic residue, h is a generator of QR_N with all but negligible probability; assume this to be the case in what follows.

Let $c \overset{\mathrm{def}}{=} a \bmod p'q'$, viewed as an integer between 0 and $p'q' - 1$ (inclusive). For any $\gamma \in \{0, \ldots, p'q' - 1\}$, the probability (over choice of a) that $c = \gamma$ is at least

$$\frac{\lfloor \frac{N^2}{p'q'} \rfloor}{N^2} \geq \frac{\frac{N^2}{p'q'} - 1}{N^2} \geq \frac{1}{p'q'} - \frac{1}{N^2}$$

and (using similar reasoning) at most $\frac{1}{p'q'} + \frac{1}{N^2}$. It follows that the distribution of c is statistically close to uniform over $\{0, \ldots, p'q' - 1\}$. Since L_{v_0} depends only on c, we conclude that the distribution of L_{v_0} is statistically close to uniform over QR_N.

We now show that when A outputs a forgery (e, auth) on a message m with $e \notin \{e_1, \ldots, e_t\}$, then A' can compute a correct solution with noticeable probability. Since $\mathrm{auth}^e = L_{v_0} h^m \bmod N$, we have

$$\mathrm{auth}^e = h^{a+m} = y^{2 \cdot (a+m) \cdot \hat{e}} \bmod N.$$

Let $K := 2 \cdot (a+m) \cdot \hat{e}$. If $\gcd(e, K) = 1$, then A' can apply Lemma 4.1 and output $(e, y^{1/e})$ as a solution. In fact, as long as e does not divide K it is possible to compute a non-trivial root: say $d = \gcd(e, K) \neq e$, and notice that — since $d < e < \min\{p', q'\}$ and e (and hence d) is odd — we have $\gcd(d, 2p'q') = 1$. Thus,

$$\mathrm{auth}^{e/d} = y^{K/d} \bmod N.$$

Now, since $\gcd(e/d, K/d) = 1$ and $e/d > 2$, algorithm A' can still apply Lemma 4.1 to compute a (non-trivial) (e/d)th root of y.

It thus only remains to show that, with noticeable probability, e does not divide K. Let r be any prime dividing[10] e, and note that $\gcd(r, 2\hat{e}) = 1$. We show that, with noticeable probability, r does not divide $(a + m)$; equivalently, we show that, with noticeable probability, $(a + m) \neq 0 \bmod r$. Write a as $a = bp'q' + c$ (so $c = a \bmod p'q'$ as before). Although L_{v_0} reveals c (in an information-theoretic sense), the conditional distribution on b given L_{v_0} is statistically close to uniform on $\{0, \ldots, \lfloor N^2/p'q' \rfloor\}$. Since $r/\lfloor N^2/p'q' \rfloor$ is negligible, it follows that $b \bmod r$ is statistically close to uniform on $\{0, \ldots, r - 1\}$ (even conditioned on the adversary's view). Using the fact that $p'q' \neq 0 \bmod r$, we see that

$$
\begin{aligned}
\Pr_b[a + m = 0 \bmod r] &= \Pr_b[(bp'q' + (c + m)) = 0 \bmod r] \\
&= \Pr_b[bp'q' = -(c + m) \bmod r] \\
&= \Pr_b[b = -(c + m)/p'q' \bmod r],
\end{aligned}
$$

and this final probability is negligibly close to $1/r \leq 1/3$. We conclude that, with noticeable probability, r does not divide $(a + m)$; hence, with noticeable probability, e does not divide K. This completes the proof.

4.3.3 The Cramer-Shoup Scheme

We now have all the building blocks in place to present the Cramer-Shoup signature scheme [36]. (See Construction 4.6. Our description is intended to highlight the correspondence with the Cramer-Damgård scheme.) The scheme assumes a collision-resistant hash function H mapping inputs to strings of length $\ell = \lfloor k/2 \rfloor$. We also continue to assume that the signature scheme is defined for messages of length ℓ.

Theorem 4.6. *If the strong RSA problem is hard relative to* GenModulus, *then the Cramer-Shoup signature scheme is existentially unforgeable under an adaptive chosen-message attack.*

Proof. All the necessary components of the proof are contained in the proofs of Theorems 4.3 and 4.5, and so we will be relatively brief here. Let A be a PPT adversary attacking the Cramer-Shoup signature scheme that outputs a forgery with probability $\varepsilon(k)$. Say A obtains $t = t(k)$ signatures $\{e_i, L_i, \text{auth}_{1,i}, \text{auth}_{2,i}\}_{i=1}^t$ on messages $\{m_i\}_{i=1}^t$, and outputs a forgery $(e, L, \text{auth}_1, \text{auth}_2)$ on a message m. Let $\varepsilon_1(k)$ denote the probability that A outputs a forgery such that $e = e_i$ for some i, and let $\varepsilon_2(k)$ be the probability that A's forgery satisfies $e \notin \{e_i\}$. Since $\varepsilon(k) = \varepsilon_1(k) + \varepsilon_2(k)$, showing that both ε_1 and ε_2 are negligible will complete the proof.

We can bound ε_1 by exactly following the proof of Theorem 4.3. In fact, we can bound ε_1 by direct reduction to the Cramer-Damgård scheme using depth $d = 1$.[11]

---

[10] Recall that signature verification does not check whether e is prime.

[11] Hardness of the strong RSA problem relative to GenModulus implies hardness of the RSA problem relative to GenModulus and an arbitrary GenPrime. This, in turn, implies security of the Cramer-Damgård scheme by theorem 4.3.

Construction 4.6: The Cramer-Shoup scheme

Let GenModulus, GenPrime be as described in the text, and set $\ell = \lfloor k/2 \rfloor$.

Key generation: On security parameter 1^k, proceed as follows:

- Run $(N, p, q) \leftarrow \mathsf{GenModulus}(1^k)$, where p and q are strong primes.
- Choose random $L_{v_0}, h \in \mathsf{QR}_N$ and compute $e' \leftarrow \mathsf{GenPrime}(1^{\ell+1})$.
- The public key is (N, L_{v_0}, h, e') and the secret key is p, q.

Signature generation: To sign message $m \in \{0,1\}^\ell$, viewed as an integer in the range $\{0, \ldots, 2^\ell - 1\}$, choose random quadratic residue $L \in \mathsf{QR}_N$ and set $e \leftarrow \mathsf{GenPrime}(1^{\ell+1})$. Then compute

$$\mathsf{auth}_1 := \left(L_{v_0} \cdot h^{H(L)} \right)^{1/e} \bmod N$$

and

$$\mathsf{auth}_2 := (L \cdot h^m)^{1/e'} \bmod N.$$

The signature is $(e, L, \mathsf{auth}_1, \mathsf{auth}_2)$.

Signature verification: To verify signature $(e, L, \mathsf{auth}_1, \mathsf{auth}_2)$ on message m, first verify that e is an odd, $(\ell+1)$-bit integer different from e' (note that it is not required to verify primality of e). Then check that

$$\mathsf{auth}_1^e \stackrel{?}{=} L_{v_0} \cdot h^{H(L)} \bmod N$$

and

$$\mathsf{auth}_2^{e'} \stackrel{?}{=} L \cdot h^m \bmod N.$$

In detail, we construct an adversary A' for the Cramer-Damgård scheme as follows: Given public key $(N, L_{v_0}, h, E = \{e_1, \ldots, e_t, e_{t+1}\})$ for an instance of the Cramer-Damgård scheme, A' runs A on public key (N, L_{v_0}, h, e_{t+1}). The t signing queries of A are answered in the natural way, and a forgery by A in which $e \in \{e_1, \ldots, e_t\}$ immediately yields a forgery for A' as well. We therefore turn our attention to bounding ε_2.

The bound on ε_2 is derived in a manner essentially the same as that used in the proof of Theorem 4.5. In particular, we can construct an algorithm A' (using A as a subroutine) that solves an instance (N, y) of the strong RSA problem with probability polynomially related to $\varepsilon_2(k)$. The public key is constructed by A' exactly as in the proof of Theorem 4.5, with the only difference being the additional selection of a prime $e' \leftarrow \mathsf{GenPrime}(1^{\ell+1})$ and its inclusion in the public key. As described in the proof of that theorem, A' will be able to compute $\mathsf{auth}_{1,i} \stackrel{\text{def}}{=} \left(L_{v_0} \cdot h^{H(L_i)} \right)^{1/e_i}$ for arbitrary L_i. To respond to the ith signature request of A on some message m_i, we thus have A' proceed as follows:

- Choose random $s_i \leftarrow \mathbb{Z}_N^*$.
- Set $L_i := s_i^{e'} \cdot h^{-m_i} \bmod N$ and $\text{auth}_{2,i} := s_i$.
- Compute $\text{auth}_{1,i} \stackrel{\text{def}}{=} \left(L_{v_0} \cdot h^{H(L_i)} \right)^{1/e_i}$ as in the proof of Theorem 4.5.
- Output the signature $(e_i, L_i, \text{auth}_{1,i}, \text{auth}_{2,i})$.

Arguing as in the proof of Theorem 4.5, we have that A will output a forgery such that $e \notin \{e_i\}$ with probability negligibly close to $\varepsilon_2(k)$. Furthermore, as in that proof, when this occurs A' can compute a non-trivial root of y with noticeable probability. It follows that ε_2 is negligible.

This completes the proof.

4.3.4 The Fischlin Scheme

An improvement to the Cramer-Shoup scheme was subsequently proposed by Fischlin [48] and is presented as Construction 4.7. This variant offers the advantages of faster signing and verification, as well as shorter signatures. It is fair to say that this scheme is currently the most efficient known scheme which can be proven secure based on standard assumptions.

Theorem 4.7. *If the strong RSA problem is hard relative to* GenModulus, *then Construction 4.7 Fischlin is existentially unforgeable under an adaptive chosen-message attack.*

Proof. The proof relies on many of the same ideas as the proof of Theorem 4.6, and also has some similarities with Construction 1.1. As usual, let A be a PPT adversary attacking the scheme that outputs a forgery with probability $\varepsilon(k)$. Say A obtains $t = t(k)$ signatures $\{(e_i, \alpha_i, \text{auth}_i)\}_{i=1}^t$ on adaptively chosen-messages $\{m_i\}_{i=1}^t$ and outputs forgery (e, α, auth) on a message m. Let $\varepsilon_1(k)$ denote the probability that A's forgery satisfies $e = e_i$ for some (unique[12]) i, and let $\varepsilon_2(k)$ be the probability that $e \notin \{e_i\}$. As in the previous proofs, we are done once we show that ε_1 and ε_2 are negligible.

If A's forgery (e, α, auth) satisfies $e = e_i$ then either $\alpha \neq \alpha_i$ or $\alpha \oplus m \neq \alpha_i \oplus m_i$ (or possibly both). Let $\varepsilon_1^1(k)$ (resp., $\varepsilon_1^2(k)$) denote the probability that the first (resp., second) case occurs. Note that $\varepsilon_1(k) \leq \varepsilon_1^1(k) + \varepsilon_1^2(k)$. We will prove that ε_1^1 is negligible; the proof that ε_1^2 is negligible is completely analogous.

We show an algorithm A_1' that solves an instance (N, y, e) of the (regular) RSA problem with probability polynomially related to $\varepsilon_1^1(k)$. A_1' begins by making a random guess $j \leftarrow \{1, \ldots, t\}$ (recall, t is the number of signatures requested by A). Next, A_1' prepares a public key as follows:

---

[12] We assume the $\{e_i\}$ are distinct, since a collision occurs with negligible probability.

Construction 4.7: The Fischlin scheme

Let GenModulus, GenPrime be as described in the text, and set $\ell = \lfloor k/2 \rfloor$.

Key generation: On security parameter 1^k, proceed as follows:

- Run $(N, p, q) \leftarrow$ GenModulus(1^k), where p and q are strong primes.
- Choose random $L, h_1, h_2 \in \mathsf{QR}_N$.
- The public key is (N, L, h_1, h_2) and the secret key is p, q.

Signature generation: To sign message $m \in \{0, 1\}^\ell$, compute $e \leftarrow$ GenPrime$(1^{\ell+1})$ and choose a random ℓ-bit string α. Then compute

$$\mathsf{auth} := \left(L \cdot h_1^\alpha \cdot h_2^{\alpha \oplus m} \right)^{1/e} \bmod N$$

(using the known factorization of N), viewing both m and $\alpha \oplus m$ as integers in the range $\{0, \ldots, 2^\ell - 1\}$. The signature is $(e, \alpha, \mathsf{auth})$.

Signature verification: To verify signature $(e, \alpha, \mathsf{auth})$ on message m, first check that e is an odd, $(\ell + 1)$-bit integer (again, we stress that verifying the primality of e is not necessary). Then check whether

$$\mathsf{auth}^e \overset{?}{=} L \cdot h_1^\alpha \cdot h_2^{\alpha \oplus m} \bmod N.$$

- Set $e_j := e$ and compute $e_j \leftarrow$ GenPrime$(1^{\ell+1})$ for $j \in \{1, \ldots, t\} \setminus \{i\}$. Then set $\hat{e} := \prod_i e_i$.
- Choose random $v, w \leftarrow \mathsf{QR}_N$ and set $h_1 := y^{2\hat{e}/e_j} \bmod N$ and $h_2 := v^{\hat{e}} \bmod N$. Choose $\alpha_j \leftarrow \{0, 1\}^\ell$ and set $L := h_1^{-\alpha_j} w^{\hat{e}} \bmod N$.
- The public key is (N, L, h_1, h_2).

The resulting public key is distributed identically to the public keys output by the true key generation algorithm. (Here, we rely on the fact that v, w are generators of QR_N except with negligible probability, something that is true because N is a product of strong primes.) Furthermore, it is fairly immediate that for $i \neq j$ it is possible for A_1' to generate a correctly distributed signature (using e_i) on any message m_i chosen by A: in particular, A_1' need only choose a random α_i and then compute

$$\begin{aligned} \mathsf{auth}_i &:= y^{2(\alpha_i - \alpha_j)\hat{e}/e_i e_j} \cdot w^{\hat{e}/e_i} \cdot v^{(\alpha \oplus m)\hat{e}/e_i} \bmod N \\ &= h_1^{(\alpha_i - \alpha_j)/e_i} \cdot w^{\hat{e}/e_i} \cdot h_2^{(\alpha \oplus m)/e_i} \bmod N \\ &= \left(L \cdot h_1^{\alpha_i} \cdot h_2^{\alpha \oplus m} \right)^{1/e_i} \bmod N. \end{aligned}$$

Finally, A_1' can also compute a signature using e_j on any message m_j chosen by A; here, A_1' uses the α_j it had chosen in advance and sets

$$\mathsf{auth}_j := w^{\hat{e}/e_j} \cdot v^{(\alpha_j \oplus m)\hat{e}/e_j} \bmod N$$

$$= \left(w^{\hat{e}} \cdot h_1^{-\alpha_j} \cdot h_1^{\alpha_j} \cdot h_2^{\alpha_j \oplus m}\right)^{1/e_j} \bmod N$$

$$= \left(L \cdot h_1^{\alpha_j} \cdot h_2^{\alpha_j \oplus m}\right)^{1/e_j} \bmod N.$$

Since α_j was chosen uniformly, and is independent of the public key, this signature too has the correct distribution.

Continuing the description of A_1': it runs A using the generated public key and answers the signing queries of A as outlined above. Assume A outputs a valid forgery $(e, \alpha, \mathsf{auth})$ on a message m with $e = e_j$ and $\alpha \neq \alpha_j$. Since A_1' perfectly simulates a real execution of the scheme and the guess j of A_1' is correct with probability $1/t$, this occurs with probability $\varepsilon_1^1(k)/t$. Moreover, when this occurs we have $\mathsf{auth}^{e_j} = L \cdot h_1^{\alpha} \cdot h_2^{\alpha \oplus m} \bmod N$ and so

$$h_1^{-\alpha_j} \cdot w^{\hat{e}} = L = \mathsf{auth}^{e_j} \cdot h_1^{-\alpha} \cdot h_2^{-(\alpha \oplus m)} \bmod N$$

or, re-writing,

$$h_1^{\alpha - \alpha_j} = \mathsf{auth}^{e_j} \cdot h_2^{-(\alpha \oplus m)} \cdot w^{-\hat{e}} \bmod N.$$

Due to the way h_1, h_2 were computed, this implies:

$$y^{2 \cdot (\alpha - \alpha_j) \cdot \hat{e}/e_j} = \mathsf{auth}^{e_j} \cdot \left(v^{\hat{e}}\right)^{-(\alpha \oplus m)} \cdot w^{-\hat{e}} \bmod N$$

$$= \left(\mathsf{auth} \cdot (v^{\hat{e}/e_j})^{-(\alpha \oplus m)} \cdot w^{-\hat{e}/e_j}\right)^{e_j} \bmod N.$$

Note that A_1' can compute $\gamma \overset{\mathrm{def}}{=} \mathsf{auth} \cdot (v^{\hat{e}/e_j})^{-(\alpha \oplus m)} \cdot w^{-\hat{e}/e_j} \bmod N$. So, A_1' can compute y^{1/e_j} by application of Lemma 4.1 (the $\{e_i\}_{i \neq j}$ are all relatively prime to e_j, and $|\alpha - \alpha_j| < e_j$ so that difference is relatively prime to e_j as well). We conclude that A_1' correctly solves its given RSA instance with probability polynomially related to ε_1^1, and so this must be negligible. (As we have previously stated, a similar argument applies to ε_1^2. Hence, this concludes the proof that ε_1 is negligible.)

We next turn to bounding ε_2. Here we construct an algorithm A_2' that solves an instance (N, y) of the strong RSA problem with probability polynomially related to $\varepsilon_2(k)$. Since the analysis here largely follows that of the proofs for Theorems 4.5 and 4.6, we merely provide a sketch. A_2' constructs a public key as follows:

- Compute $e_i \leftarrow \mathsf{GenPrime}(1^{\ell+1})$ for $i = 1, \ldots, t$. Then set $\hat{e} := \prod_i e_i$
- Set $h_1 := y^{2\hat{e}} \bmod N$. Choose $a, a' \leftarrow \{1, \ldots, N^2\}$, and set $h_2 := h_1^a \bmod N$ and $L := h_1^{a'} \bmod N$.
- The public key is (N, L, h_1, h_2).

As argued in the proof of Theorem 4.5, the distribution on the public key (N, L, h_1, h_2) generated in this way is statistically close to the distribution on public keys output by the real key generation procedure. Furthermore, since A_2' knows the e_i^{th} roots of L, h_1, and h_2 for all i, it is not hard to see that A_2' can provide any signatures requested by A (and these signatures will have the correct distribution). Finally, if A outputs a valid forgery $(e, \alpha, \mathsf{auth})$ on some message m, and it is the case that $e \notin \{e_i\}$, then

$$\mathsf{auth}^e = L \cdot h_1^\alpha \cdot h_2^{\alpha \oplus m} \bmod N$$
$$= y^{2\hat{e} \cdot (a' + \alpha + a \cdot (\alpha \oplus m))} \bmod N$$

from which A_2' can compute a non-trivial root of y with noticeable probability (as in the proof of Theorem 4.5). This concludes the proof.

4.3.5 The Gennaro-Halevi-Rabin Scheme

The scheme presented here as Construction 4.8, due to Gennari, Halevi, and Rabin [53], is markedly different from the Cramer-Shoup-Fischlin scheme, and can be said to more directly follow the ideas of Lemma 4.2. As presented here, the Gennaro-Halevi-Rabin scheme has the advantage of being very easy to analyze although it is less efficient than the previous two schemes we have seen. The original paper discusses a more efficient variant of the scheme that is more difficult to analyze, and relies on a somewhat non-standard assumption.

For the purposes of the presentation, we assume there exists an efficient, deterministic function f mapping messages to odd primes of length at least k. (This assumption on f can be relaxed [53] at the expense of a more complicated analysis. Again, our intention here is not to present the most optimized instantiation of the scheme, but only to convey the main ideas.) Furthermore, f is required to be collision resistant. We refer the reader to the bibliographic notes at the end of this chapter for references to various constructions of such an f.

Construction 4.8: The Gennaro-Halevi-Rabin scheme

Let GenModulus and f be as described in the text.

Key generation: On security parameter 1^k, proceed as follows:

- Run $(N, p, q) \leftarrow \mathsf{GenModulus}(1^k)$ with p, q strong primes.
- Choose random $z \leftarrow \mathbb{Z}_N^*$.
- The public key is (N, z) and the secret key is p, q.

Signature generation: The signature on message m is $z^{1/f(m)} \bmod N$, computed using the factorization of N.

Signature verification: To verify signature σ on message m, simply output 1 iff $\sigma^{f(m)} \overset{?}{=} z \bmod N$.

Note the strong parallel with what is claimed in Lemma 4.2. A key difference is that here the set of roots $\{z^{1/f(m_i)} \bmod N\}$ that are available to an adversary may be chosen adaptively since the primes $\{f(m_i)\}$ depend on message $\{m_i\}$ that are potentially under the control of an adversary. Indeed, it is not known whether the scheme as presented in Construction 4.8 can be proven secure against adaptive chosen-message attacks based on the strong RSA assumption. (Although we remark that security against adaptive chosen-message attacks can be proved based on a somewhat non-standard assumption [53].) Nevertheless, we can prove the scheme secure against *known* message attacks.

Theorem 4.8. *If the strong RSA problem is hard relative to* GenModulus, *the Gennaro-Halevi-Rabin scheme is strongly unforgeable under a known-message attack.*

Proof. Every message has a unique valid signature, so existential unforgeability and strong unforgeability and it suffices to prove the former. Given a PPT adversary A succeeding in a known-message attack with probability $\varepsilon(k)$, we construct an algorithm A' that solves a given instance (N, y) of the strong RSA problem with the same probability. It follows by assumption that $\varepsilon(k)$ must be negligible.

The attack begins when $A(1^k)$ outputs a set of $t = t(k)$ messages $\{m_i\}_{i=1}^t$. Next, A' prepares a public key and signatures as follows: it computes $\hat{e} := \prod_i f(m_i)$ and sets $z := y^{\hat{e}} \bmod N$. The public key is (N, z). Furthermore, the signature on each message m_i is computed as

$$\sigma_i := y^{\hat{e}/f(m_i)} = z^{1/f(m_i)} \bmod N$$

(note that A' can compute this value since $\hat{e}/f(m_i)$ is an integer). It is easy to see that the public key and all signatures are distributed exactly as in an execution of the real signature scheme.

Now, if A outputs a valid forgery (m, σ) with $m \notin \{m_i\}$ then we have

$$\sigma^{f(m)} = z = y^{\hat{e}} \bmod N.$$

Since $f(m)$ is prime and f is collision resistant, we have $\gcd(f(m), \hat{e}) = 1$ (except with negligible probability) and A' can compute a non-trivial $f(m)$th root of y using Lemma 4.1. This concludes the proof.

4.4 Further Reading

It is interesting to observe that all known schemes based on the RSA assumption with even moderate efficiency require the RSA problem (i.e., computing $y^{1/e} \bmod N$) to be hard for several primes e (e.g., a random large prime as in the constructions described here, or several fixed small primes). It remains open to construct an efficient scheme based on the assumption, say, that the RSA problem with fixed exponent $e = 3$ is hard.

Extending the ideas used in the Dwork-Naor [42] and Cramer-Damgård [34] signature schemes, Catalano and Gennaro [26] show a tree-based signature scheme based on a variant of the factoring assumption.

The strong RSA assumption was introduced in [50, 4]. Although seemingly much stronger than the standard RSA assumption, the best known attacks on the RSA and strong RSA problems currently have the same complexity.

Construction 4.8 is based on the original Gennaro-Halevi-Rabin signature scheme introduced in [53], and was also implicitly considered by Naccache, et al. [86]. In the original paper presenting the scheme [53], existential unforgeability is shown directly (i.e., without resorting to Theorem 1.1) by introducing slight modifications to the scheme and imposing additional assumptions on the function f (but see [31] for further discussion regarding the validity of those assumptions). Functions f which unconditionally satisfy the properties as required for Theorem 4.8 are given in [23, 83, 31, 86]; none of these constructions, however, is particularly efficient.

Besides the schemes discussed in this chapter, other notable signature schemes based on the strong RSA assumption includes [24, 75, 27, 65]; see [70] for a recent survey of those and other works.

As we have mentioned, the signature schemes of Cramer-Shoup and Fischlin are currently the most efficient provably secure schemes known. In our description of these schemes, however, we have been concerned more with clarity than efficiency. We refer the reader to [36, 48] for various efficiency improvements as well as discussions on efficient implementation of these schemes.

A second paper by Hohenberger-Waters [66] shows an approach by which certain signature schemes based on the strong RSA assumption can be converted to *stateful* schemes based on the standard RSA assumption.

Chapter 5
Constructions Based on Bilinear Maps

5.1 Introduction

In the past 10 years cryptographic constructions based on *bilinear maps* have become extremely popular, most prominently following their use in constructing identity-based encryption schemes. Bilinear maps have also led to several efficient signature schemes, and we explore two such constructions here.

5.1.1 Technical Preliminaries

For the purposes of this chapter, we treat bilinear maps in a completely abstract fashion without discussing any specific instantiation. A detailed understanding of bilinear maps as used in cryptography requires a more in-depth familiarity with elliptic curves than we are willing to assume on behalf of the reader. The references at the end of this chapter serve as a good starting point for further reading on this topic.

Let $\mathcal{G}$ be an efficient algorithm that, on input 1^k, outputs $(\mathbb{G}, \mathbb{G}_T, q, g, \hat{e})$ where:

- $\mathbb{G}, \mathbb{G}_T$ are (descriptions of) two groups of prime order q, where group operations in $\mathbb{G}, \mathbb{G}_T$ can be performed efficiently.
- g is a random[1] generator of $\mathbb{G}$.
- $\hat{e}$ is a (description of an) efficiently computable function[2] $\hat{e} : \mathbb{G} \times \mathbb{G} \to \mathbb{G}_T$ whose properties are summarized below.

---

[1] The assumption that g is random can be relaxed, but assuming g is random simplifies things.

[2] Often $\hat{e}$ is defined as a mapping $\hat{e} : \mathbb{G}_1 \times \mathbb{G}_2 \to \mathbb{G}_T$ where $\mathbb{G}_1, \mathbb{G}_2$ are isomorphic but do not necessarily share the same representation. (This is sometimes referred to as the *asymmetric* case, in contrast to the *symmetric* case presented in this text.) We have opted to focus on the symmetric case because doing so allows us to simplify the notation; all the results covered in this chapter, however, carry over to the asymmetric setting.

J. Katz, *Digital Signatures*, DOI 10.1007/978-0-387-27712-7_5,
© Springer Science+Business Media, LLC 2010

The function $\hat{e}$ is required to be *bilinear*, meaning that for every $g, h \in \mathbb{G}$ and every $a, b \in \mathbb{Z}_q$

$$\hat{e}(g^a, h^b) = \hat{e}(g, h)^{ab}.$$

We also assume that $\hat{e}$ is *non-degenerate*, namely that $\hat{e}(g,g) \neq 1$.

Later, we will formally state the specific computational assumptions that will be used to prove security of the signature schemes that will be presented. To gain some familiarity with bilinear maps, though, we include some preliminary discussion here.

We always assume that the discrete logarithm problem is hard in $\mathbb{G}$. Formally, we assume the following is negligible for all PPT algorithms A:

$$\Pr[(\mathbb{G}, \mathbb{G}_T, q, g, \hat{e}) \leftarrow \mathcal{G}(1^k); h \leftarrow \mathbb{G}; x \leftarrow A(1^k, \mathbb{G}, \mathbb{G}_T, q, g, h, \hat{e}) : g^x = h].$$

This implies hardness of the discrete logarithm problem in $\mathbb{G}_T$: if not, then A could efficiently compute the discrete logarithm of $h \in \mathbb{G}$ as follows:

- Compute $g_T = \hat{e}(g, g)$. By non-degeneracy of $\hat{e}$ and the fact that the orders of $\mathbb{G}$ and $\mathbb{G}_T$ are prime, we have that g_T is a generator of $\mathbb{G}_T$.
- Compute $h_T = \hat{e}(g, h)$.
- Compute $\log_{g_T} h_T = x$ (using the fact that the discrete logarithm problem in $\mathbb{G}_T$ is easy).

This succeeds because of bilinearity; namely, because if $h = g^x$ then

$$h_T = \hat{e}(g, h) = \hat{e}(g, g^x) = \hat{e}(g, g)^x = g_T^x.$$

5.1.2 Outline of the Chapter

In this chapter we focus on two signature schemes built using bilinear maps. The first scheme we present is due to Boneh and Boyen and relies on an assumption similar in some respects to the strong RSA assumption. We then show a scheme due to Waters that can be based on the more standard computational Diffie-Hellman assumption (though in a group where an efficiently computable bilinear map is defined). This scheme is interesting in that it is derived from an identity-based encryption scheme; the reader is referred to Waters's original paper [109] for further details.

Our intention here is to highlight the techniques used in designing these schemes and proving them secure, not to discuss their relative performance benefits. Such a discussion would require a more extensive digression into the specifics of bilinear maps as used in cryptography.

5.2 The Boneh-Boyen Scheme

Let $\mathcal{G}$, as before, be an algorithm that generates parameters $(\mathbb{G}, \mathbb{G}_T, q, g, \hat{e})$ defining a bilinear map. For compactness, we abbreviate the output of $\mathcal{G}$ by params. The *strong Diffie-Hellman* (SDH) assumption states that the following problem is hard:

> Given params and $g^x, g^{x^2}, \ldots, g^{x^q} \in \mathbb{G}$ (for random x),
> output $\left(c, g^{1/(x+c)}\right)$ for an arbitrary $c \in \mathbb{Z}_q$.

(The solution with $c = -x$ is disallowed.) We will in fact require this problem to be hard for *any* (polynomial) q. Observe that a candidate solution $(c, h) \in \mathbb{Z}_q \times \mathbb{G}$ can be verified (with respect to the given instance $g, g^x, \ldots, g^{x^q}$) by checking whether $\hat{e}(g^x g^c, h) \stackrel{?}{=} \hat{e}(g, g)$. Letting $h = g^{1/(x+c')}$ for some $c' \in \mathbb{Z}_q$ (note that if $h = 1$ then $\hat{e}(g^x g^c, h) = 1 \neq \hat{e}(g, g)$; in any other case, c' is well-defined), this works because

$$\hat{e}(g^x g^c, h) = \hat{e}(g^{x+c}, g^{1/(x+c')}) = \hat{e}(g, g)^{(x+c)/(x+c')},$$

and the final expression is equal to $e(g, g)$ if and only if $c = c'$.

Formally:

Definition 5.1. The **SDH problem is hard relative to** $\mathcal{G}$ if for all PPT algorithms A and all polynomials $q = q(k)$ the following is negligible:

$$\Pr\left[\begin{array}{l} \text{params} \leftarrow \mathcal{G}(1^k); x \leftarrow \mathbb{Z}_q; \\ (c, h) \leftarrow A(\text{params}, g^x, g^{x^2}, \ldots, g^{x^q}) \end{array} : \hat{e}(g^x g^c, h) = \hat{e}(g, g)\right].$$

The SDH assumption, as stated, is vaguely similar to the strong RSA assumption insofar as A is given the freedom to choose c. A stronger correspondence is given by the following lemma, which can be viewed as an analogue of Lemma 4.2.

Lemma 5.1. *Say the SDH problem is hard relative to $\mathcal{G}$. Then for all PPT algorithms A, all polynomials q, and any set of integers $\{c_i\}_{i=1}^q$, the following is negligible:*

$$\Pr\left[\begin{array}{l} \text{params} \leftarrow \mathcal{G}(1^k); x \leftarrow \mathbb{Z}_q; \\ (c, h) \leftarrow A\left(\text{params}, g^x, \{c_i, g^{1/(x+c_i)}\}_{i=1}^q\right) \end{array} : \hat{e}(g^x g^c, h) = \hat{e}(g, g) \bigwedge c \notin \{c_i\}\right].$$

Stated differently, even given several SDH solutions $\{c_i, g^{1/(x+c_i)}\}_{i=1}^q$ it remains difficult for A to find a *new* SDH solution $(c, g^{1/(x+c)})$. This immediately yields a signature scheme secure under a known-message attack, as we will see subsequently. We first prove the lemma.

Proof. Given a PPT algorithm A, a polynomial q, and a set of integers $\{c_i\}_{i=1}^q$ as in the lemma, we construct a PPT algorithm B solving the SDH problem. Algorithm B is given params $= (\mathbb{G}, \mathbb{G}_T, q, g, \hat{e})$ and $\{g^{x^i}\}_{i=1}^{q+1}$. A key observation is that given any polynomial $f(X) \in \mathbb{Z}_q[X]$ of degree at most $q+1$, it is possible for B to compute

$g^{f(x)}$ (where x here is the same as in the instance B is given). This is done by writing $f(X) = \sum_{i=0}^{q+1} f_i X^i$ and then computing

$$\prod_{i=0}^{q+1} \left(g^{x^i} \right)^{f_i} = g^{\sum_{i=0}^{q+1} f_i x^i}.$$

We use this extensively in what follows.

B defines the polynomial $f(X) = \prod_{i=1}^{q}(X + c_i)$ and chooses a random $r \leftarrow \mathbb{Z}_q$. It then computes the following values, exactly as described above:

1. $\bar{g} := \left(g^{f(x)} \right)^r$.

2. $\bar{g}' := \left(g^{xf(x)} \right)^r$. (Note that the polynomial $Xf(X)$ has degree $q+1$.)

3. For $i \in \{1, \ldots, q\}$, $h_i := \left(g^{f(x)/(x+c_i)} \right)^r$. (Note that, by construction of f, we indeed have that $f(X)/(X + c_i) \in \mathbb{Z}_q[X]$ for all i.)

B then runs A on input params$' = (\mathbb{G}, \mathbb{G}_T, q, \bar{g}, \hat{e})$, $\bar{g}'$, and $\{c_i, h_i\}_{i=1}^{q}$. Observe that

- $\bar{g}$ is a random element of $\mathbb{G}$;
- $\bar{g}' = \bar{g}^x$ for a random $x \in \mathbb{Z}_q$ (unknown to B);
- $h_i = \bar{g}^{1/(x+c_i)}$ for all i.

We assume $f(x) \neq 0$ since this event occurs with negligible probability (and would anyway allow B to compute x). Thus, the inputs given to A have the correct distribution and so the probability with which A outputs $(c, \bar{h})$ with $\bar{h} = \bar{g}^{1/(x+c)}$ and $c \notin \{c_i\}$ remains unchanged.

Assuming A does indeed output $(c, \bar{h})$ with $\bar{h} = \bar{g}^{1/(x+c)}$ and $c \notin \{c_i\}$, we show that it is possible for B to compute a valid solution to its SDH instance. Using long division, B computes a polynomial $f' \in \mathbb{Z}_q[X]$ and a non-zero scalar $c' \in \mathbb{Z}_q$ such that $f(X) = f'(X) \cdot (X + c) + c'$. (We have $c' \neq 0$ by definition of f and the fact that $c \notin \{c_i\}$.) Then B outputs the solution

$$\left(c, \left(\bar{h}^{1/r} / g^{f'(x)} \right)^{1/c'} \right),$$

where the value $g^{f'(x)}$ is computed as described earlier. This is a valid solution to B's given SDH instance since

$$\left(\frac{\bar{h}^{1/r}}{g^{f'(x)}} \right)^{1/c'} = \left(\frac{\bar{g}^{1/(r(x+c))}}{g^{f'(x)}} \right)^{1/c'}$$

$$= \left(\frac{g^{rf(x)/(r(x+c))}}{g^{f'(x)}} \right)^{1/c'}$$

$$= \left(\frac{g^{f'(x)+\frac{c'}{x+c}}}{g^{f'(x)}} \right)^{1/c'} = \left(g^{\frac{c'}{x+c}} \right)^{1/c'} = g^{1/(x+c)}.$$

The previous lemma suggests the following signature scheme. The public key contains $(\mathbb{G}, \mathbb{G}_T, q, g, \hat{e})$ along with a value $X = g^x$; the secret key is x. The signature on the message $m \in \mathbb{Z}_q$ is just $\sigma = g^{1/(x+m)}$. (We require $m \neq -x$.) To verify a signature σ on a message m, it is simply required to check whether

$$\hat{e}(X\, g^m, \sigma) \overset{?}{=} \hat{e}(g, g);$$

this succeeds because for a correct signature σ we have

$$\hat{e}(X\, g^m, \sigma) = \hat{e}(g^x g^m, g^{1/(x+m)}) = \hat{e}(g, g)^{(x+m)/(x+m)} = \hat{e}(g, g).$$

Lemma 5.1 immediately implies that the scheme just described is existentially unforgeable under a known-message attack. We could of course then apply Construction 1.2 to obtain a scheme secure under an adaptive chosen-message attack. The Boneh-Boyen scheme, given as Construction 5.1, applies essentially the same idea but more efficiently.

Construction 5.1: The Boneh-Boyen scheme

Let $\mathscr{G}$ be as described in the text.

Key generation: Compute params $\overset{\text{def}}{=} (\mathbb{G}, \mathbb{G}_T, q, g, \hat{e}) \leftarrow \mathscr{G}(1^k)$ and choose random $x, y \in \mathbb{Z}_q$. The public key is $(\text{params}, g^x, g^y)$ and the secret key is x, y.

Signature generation: Given a message $m \in \mathbb{Z}_q$, the signer choose a random $r \in \mathbb{Z}_q$ and outputs the signature $\left(r, g^{1/(x+m+yr)}\right)$. (Note that $x + m + yr \neq 0 \bmod q$ except with negligible probability over choice of r.)

Signature verification: To verify a signature (r, σ) on a message m with respect to a public key (params, X, Y), check whether $\hat{e}(X g^m Y^r, \sigma) \overset{?}{=} \hat{e}(g, g)$.

Theorem 5.1. *If the SDH problem is hard relative to $\mathscr{G}$, then the Boneh-Boyen scheme is strongly unforgeable under an adaptive chosen-message attack.*

Proof. Fix some PPT algorithm A attacking the signature scheme, and let q be a (polynomial) upper-bound on the number of signature queries made by A. For a given public key (params, X, Y) (with associated secret key x, y), say A obtains signatures $\{(r_i, \sigma_i)\}$ on messages $\{m_i\}$ and outputs a strong forgery $m, (r, \sigma)$. (We allow $m \in \{m_i\}$ as long as $(m, r, \sigma) \notin \{(m_i, r_i, \sigma_i)\}$.) We assume A never requests a signature on the message $m = -x$; this is without loss of generality since, given $m = -x$, it is trivial to construct a forgery. Let $\varepsilon_1(k)$ denote the probability with which A outputs a valid forgery and $m + ry = m_i + r_i y$ for some i, and let $\varepsilon_2(k)$ denote the probability with which A outputs a forgery and $m + ry \neq m_i + r_i y$ for all i. (Note that one can efficiency determine which is the case, even given only the public key, by checking if $g^m Y^r = g^{m_i} Y^{r_i}$ for some i.) We show that ε_1 and ε_2 are negligible; since the overall success probability of A is $\varepsilon(k) = \varepsilon_1(k) + \varepsilon_2(k)$, this completes the proof.

We begin by bounding ε_2 since this is the easier case. Consider the following algorithm B_2 given input as in Lemma 5.1 for *random* $\{c_i\}_{i=1}^q$:

> **Algorithm B_2:**
> The algorithm chooses[3] random $\{c_i\}_{i=1}^q$, and is then given params, $g^x, \{c_i, \sigma_i = g^{1/(x+c_i)}\}_{i=1}^q$.
>
> - Choose $y \leftarrow \mathbb{Z}_q \setminus \{0\}$, and then give to A the public key (params, g^x, g^y).
> - When A requests a signature on the ith message m_i, set $r_i := (c_i - m_i)/y \bmod q$ and return the signature (r_i, σ_i).
> - If A outputs a strong forgery $m, (r, \sigma)$ with $m + ry \neq m_i + r_i y$ for all i, then output the solution $(m + ry, \sigma)$.

Since the $\{c_i\}$ are chosen at random, independent of the public key given to A, the distribution over A's view in the above experiment is statistically close to the distribution over A's view in a real attack on the signature scheme. (In particular, the $\{r_i\}$ are random and signatures constructed by B_2 are correctly distributed.) Thus, with probability exactly ε_2 in the above experiment, A outputs a strong forgery $m, (r, \sigma)$ where $m + ry \neq m_i + r_i y$ for all i. In this case, $\sigma = g^{1/(x+(m+ry))}$. Further, by construction of B_2 we have $c_i = m_i + r_i y$ for all i. Thus, $m + ry \notin \{c_i\}$ and therefore B_2 outputs a *new* (and correct) SDH solution with probability ε_2. By the assumed hardness of the SDH problem and Lemma 5.1, it follows that ε_2 is negligible.

We now bound ε_1. The main idea is similar to that used above, but the algebra is slightly more complicated. Consider the following algorithm B_1

> **Algorithm B_1:**
> The algorithm chooses random non-zero $\{c_i\}_{i=1}^q$, and is then given params, $g^y, \{c_i, \sigma_i = g^{1/(y+c_i)}\}_{i=1}^q$.
>
> - Choose $x \leftarrow \mathbb{Z}_q$, and give the public key (params, g^x, g^y) to A.
> - When A requests a signature on the ith message m_i, set $r_i := (x + m_i)/c_i \bmod q$ and return the signature $\left(g^{1/(y+c_i)}\right)^{1/r_i}$. (Note that $r_i \neq 0$ by our assumption that $m_i \neq -x$.)
> - If A outputs a strong forgery (r, σ) on a message m, with $g^m (g^y)^r = g^{m_i}(g^y)^{r_i}$ for some i, then compute
>
> $$y := (m - m_i)/(r_i - r) \bmod q.$$
>
> (We show below that $r \neq r_i$.) At this point it is trivial to output a new SDH solution.

---

[3] Although Lemma 5.1 was only stated for the case of fixed $\{c_i\}$, it is not difficult to see that the lemma extends to this setting as well.

It is straightforward to see that the public key given to A has the same distribution as in a real attack, and the $\{r_i\}$ are all random. It takes a little more work to see that the signatures constructed by B_1 have the correct form; this is so because

$$\left(g^{1/(y+c_i)}\right)^{1/r_i} = g^{1/(r_iy+r_ic_i)} = g^{1/(r_iy+x+m_i)},$$

by choice of r_i. Finally, say A outputs a strong forgery $m,(r,\sigma)$ with $g^m(g^y)^r = g^{m_i}(g^y)^{r_i}$ for some i. For a given public key and m,r, there is a unique σ such that (r,σ) is a valid signature on m. Since (r,σ) is a valid signature on m but $(m,r,\sigma) \neq (m_i,r_i,\sigma_i)$, we must in fact have $(m,r) \neq (m_i,r_i)$. But if $g^m(g^y)^r = g^{m_i}(g^y)^{r_i}$ then $m+ry = m_i+r_iy$ and so actually $m \neq m_i$ and $r \neq r_i$. It follows that B_1 can correctly solve for y as described, and so B_1 outputs a *new* (and correct) SDH solution with probability ε_1. By the assumed hardness of the SDH problem and Lemma 5.1, it follows that ε_1 is negligible.

This concludes the proof.

5.3 The Waters Scheme

We conclude this chapter by presenting the Waters signature scheme. This construction is of significant interest for several *other* applications it has beyond merely serving as a standard signature scheme, most prominently to identity-based encryption (the context in which it was first proposed); unfortunately, we will not be able to survey these diverse applications in this book. For our purposes, the scheme serves as an example of an *efficient* scheme based on (what is for the most part) a *standard* assumption. The proof technique used to prove it secure is quite clever too.

We can define the *computational Diffie-Hellman* (CDH) assumption in any (cyclic) group $\mathbb{G}$, and this assumption had been used well before bilinear maps were introduced to cryptography. Roughly, the CDH assumption is that the following problem is hard for any generator $g \in \mathbb{G}$:

Given $g^a, g^b \in \mathbb{G}$ (for random exponents a,b), output $g^{ab} \in \mathbb{G}$.

Interestingly, in the general case a valid solution to the CDH problem cannot necessarily be verified; that is, there is not necessarily any efficient mechanism to determine whether a candidate solution h to a CDH instance (g, g^a, g^b) is correct or not. Moreover, in certain groups $\mathbb{G}$ it is even reasonable to assume that it is (essentially) *never* possible to verify a solution; this is, roughly speaking, what the *decisional* Diffie-Hellman assumption entails.

When an efficiently computable bilinear map is defined over $\mathbb{G}$, however, it *is* possible to efficiently verify correct solutions to the CDH problem (and so, in particular, the decisional Diffie-Hellman assumption cannot hold). Indeed, given four elements g, g^x, g^y, g^z, where g has order q, checking whether $z \stackrel{?}{=} xy \bmod q$ is equivalent to checking whether $\hat{e}(g, g^z) \stackrel{?}{=} \hat{e}(g^x, g^y)$. Assuming the CDH assumption holds, this asymmetry between the difficulty of computing a correct solution and the ease of verifying it proves very useful; we will see this both here, as well as later in Section 7.1.1.

The CDH assumption is formally stated as follows:

Definition 5.2. The **CDH problem is hard relative to** $\mathscr{G}$ if for all PPT algorithms A the following is negligible:

$$\Pr\left[\text{params} \leftarrow \mathscr{G}(1^k); a, b \leftarrow \mathbb{Z}_q : A(\text{params}, g^a, g^b) = g^{ab}\right].$$

We jump right in with a description of the Waters scheme.

Construction 5.2: The Waters scheme

Let $\mathscr{G}$ be as described in the text.

Key generation: Compute params $\stackrel{\text{def}}{=} (\mathbb{G}, \mathbb{G}_T, q, g, \hat{e}) \leftarrow \mathscr{G}(1^k)$. Choose $a \leftarrow \mathbb{Z}_q$ and set $g_1 := g^a$. Choose also random $g_2, u_0, \ldots, u_k \leftarrow \mathbb{G}$. The public key is $(\text{params}, g_1, g_2, u_0, \ldots, u_k)$ and the secret key is g_2^a.

Fixing a public key, we set $H(m) \stackrel{\text{def}}{=} u_0 \cdot \prod_{i=1}^{k} u_i^{m_i}$ where $m = m_1 \cdots m_k \in \{0,1\}^k$.

Signature generation: Given a message $m \in \{0,1\}^k$, the signer choose a random $r \in \mathbb{Z}_q$ and outputs the signature $(g_2^a \cdot H(m)^r, g^r)$.

Signature verification: To verify a signature (σ_1, σ_2) on a message m, check whether $\hat{e}(g, \sigma_1) \stackrel{?}{=} \hat{e}(\sigma_2, H(m)) \cdot \hat{e}(g_1, g_2)$.

Let us first verify correctness. Given an honestly generated signature (σ_1, σ_2) on a message m, we have

$$\begin{aligned}
\hat{e}(g, \sigma_1) &= \hat{e}(g, g_2^a \cdot H(m)^r) \\
&= \hat{e}(g, g_2^a) \cdot \hat{e}(g, H(m)^r) \\
&= \hat{e}(g^a, g_2) \cdot \hat{e}(g^r, H(m)) \\
&= \hat{e}(g_1, g_2) \cdot \hat{e}(\sigma_2, H(m)),
\end{aligned}$$

as desired.

Theorem 5.2. *If the CDH problem is hard relative to $\mathscr{G}$, then the Waters scheme is unforgeable under an adaptive chosen-message attack.*

We remark that the Waters scheme is *not* strongly unforgeable, as there is an easy way to "re-randomize" a valid signature on any message. This can actually be a useful feature for certain applications.

Proof. Given an adversary A attacking the signature scheme, making at most q_s signing queries, and having success probability ε, we construct an adversary B solving the CDH problem with probability polynomially related to ε. Informally, B will generate a public key (having the appropriate distribution) for which B can compute valid signatures on a (random) subset $S \subset \{0,1\}^k$ of messages, and such that any forgery produced by A on a message *outside* of S allows B to solve its given CDH instance. We expect B to be successful with probability roughly

$$\varepsilon \cdot \left(\frac{|S|}{2^k}\right)^{q_s} \cdot \left(1 - \frac{|S|}{2^k}\right);$$

setting $|S|/2^k \approx \left(1 - \frac{1}{q_s}\right)$ then gives B success probability $O(\varepsilon)$. The formal analysis is a bit more subtle, and is given next.

Algorithm B:
The algorithm is given $\mathsf{params} = (\mathbb{G}, \mathbb{G}_T, q, g, \hat{e})$ and g_1, g_2.

- Set $\ell = 2q_s$. We assume $k\ell < q$, which holds for k large enough.
- Choose $x_0 \leftarrow \{-k\ell, \dots, 0\}$ and $x_1, \dots, x_k \leftarrow \{0, \dots, \ell\}$. Also choose $y_0, \dots, y_k \leftarrow \mathbb{Z}_q$.
- For $i = 0, \dots, k$, set $u_i := g_2^{x_i} g^{y_i}$. Define the functions $F(m) \overset{\text{def}}{=} x_0 + \sum_{i=1}^k m_i x_i$ and $G(m) \overset{\text{def}}{=} y_0 + \sum_{i=1}^k m_i y_i$.
- Run A on the public key $(\mathsf{params}, g_1, g_2, u_0, \dots, u_k)$. When A requests a signature on a message m, do:
  1. If $F \overset{\text{def}}{=} F(m) = 0 \bmod q$ then abort.
  2. Otherwise, set $w := F^{-1} \bmod q$ and $G \overset{\text{def}}{=} G(m)$. Choose $r \leftarrow \mathbb{Z}_q$ and return the signature $\left(g_2^{Fr} g^{Gr} g_1^{-Gw}, g^r g_1^{-w}\right)$.
- If A outputs a valid forgery $(m, (\sigma_1, \sigma_2))$, do:
  1. If $F(m) \neq 0 \bmod q$ then abort.
  2. Otherwise, output $\sigma_1 / \sigma_2^{G(m)}$.

It is easy to see that the public key given to A has the correct distribution. Let us next argue that as long as B does not abort (1) the signatures given to A are distributed correctly, and (2) B outputs a correct solution to its given CDH instance. Let $a = \log_g g_1$ and $b = \log_g g_2$. Recall that a real signature on the message m would be computed as $\left(g_2^a H(m)^{\bar{r}}, g^{\bar{r}}\right)$ for a random $\bar{r}$. Setting $F = F(m)$, $G = G(m)$, $w = F^{-1} \bmod q$ (using the fact that B only returns a signature if $F \neq 0 \bmod q$), and $\bar{r} = r - aw$, and observing that $H(m) = g_2^{F(m)} g^{G(m)}$, we have

$$g_2^a H(m)^{\bar{r}} = g_2^a H(m)^{r-aw} = g_2^a \cdot \left(g_2^F g^G\right)^{r-aw}$$
$$= g_2^a g_2^{Fr-a} g^{Gr-Gaw}$$
$$= g_2^{Fr} g^{Gr} g_1^{-Gw}$$

and

$$g^{\bar{r}} = g^{r-aw} = g^r g^{-aw} = g^r g_1^{-w},$$

exactly as returned by B. Since B chooses r at random, $\bar{r} = r - aw$ is uniformly distributed and we conclude that signatures returned by B have the correct distribution.

Say A outputs a valid signature (σ_1, σ_2) on a message m for which $F(m) = 0 \bmod q$. By definition of the verification algorithm, this means $\hat{e}(g, \sigma_1)/\hat{e}(\sigma_2, H(m)) = \hat{e}(g_1, g_2) = \hat{e}(g, g^{ab})$, and so

$$\hat{e}(g,g)^{ab} = \frac{\hat{e}(g,\sigma_1)}{\hat{e}(\sigma_2, H(m))}$$

$$= \frac{\hat{e}(g,\sigma_1)}{\hat{e}(g, g_2^0 g^{G(m)})},$$

from which we conclude that $\sigma_1/g^{G(m)} = g^{ab}$.

It remains only to analyze the probability with which B does not abort, and successfully completes the simulation. Since the $\{x_i\}$ chosen by B are independent of the public key given to A, we can analyze this probability by imagining a "mental experiment" in which A interacts with the real signature scheme, requesting signatures on messages $m_1, \ldots, m_{q_s}$ and then outputting a forgery on a message $m \notin \{m_1, \ldots, m_{q_s}\}$ with (overall) probability ε. After this point, $\{x_i\}$ are chosen from the same distribution used by B, and we ask whether

$$F(m_1) \neq 0 \bmod q \ \wedge \ \cdots \ \wedge \ F(m_{q_s}) \neq 0 \bmod q \ \wedge \ F(m) = 0 \bmod q. \qquad (5.1)$$

We stress that the probability that the above event holds is *not* independent of m and the $\{m_i\}$, and this dependence creates problems some application of the Waters scheme (most notably in the context of proving that it gives a secure identity-based encryption scheme). In our context, however, it suffices to give a lower bound for the probability of the above event that holds for any m and $\{m_i\}$.

Fix arbitrary m and $\{m_i\}$, and let E be the event in Equation (5.1). We are interested in the probability of this event over choice of x and the $\{x_i\}$. We can simplify things by noting that $|F(m)| < q$ for all m, and therefore $F(m) = 0 \bmod q$ iff $F(m) = 0$. We then have

$$\Pr[\overline{E}] \leq \Pr[F(m) \neq 0] + \sum_{i=1}^{q_s} \Pr\left[F(m) = 0 \bigwedge F(m_i) = 0\right].$$

For any $x_1, \ldots, x_k$, there is exactly one choice of x_0 for which $F(m) = 0$; thus, $\Pr[F(m) \neq 0] = k\ell/(k\ell + 1)$. Next consider any $i \in \{1, \ldots, q_s\}$, and let $M = m_i$. Since $m \neq M$, there must be some index j where they differ. Without loss of generality, say the jth bit of m is 1 and the jth bit of M is 0. Fixing some choice of $x_1, \ldots, x_{j-1}, x_{j+1}, \ldots, x_k$, we see that $F(m) = 0$ and $F(M) = 0$ only if

$$x_0 + x_j = -\sum_{i \neq j} m_i x_i \quad \text{and} \quad x_0 = -\sum_{i \neq j} M_i x_i,$$

and therefore

$$\Pr[F(m) = 0 \bigwedge F(m_i) = 0] \leq \frac{1}{k\ell+1} \cdot \frac{1}{\ell+1}.$$

Putting everything together, we have

$$\Pr[\overline{E}] \leq \frac{k\ell}{k\ell+1} + \frac{q_s}{(k\ell+1)\cdot(\ell+1)},$$

and so

$$\Pr[E] \geq \frac{1}{k\ell+1} \cdot \left(1 - \frac{q_s}{\ell+1}\right) \geq \frac{1}{4kq_s+2}.$$

From this we conclude that B succeeds with probability at least $\varepsilon/(4kq_s+2)$, and so ε must be negligible if the CDH problem is hard relative to $\mathcal{G}$.

5.4 Further Reading

A good introduction to elliptic-curve cryptography, including a discussion of bilinear maps, can be found in the textbook by Washington [108]. Galbraith et al. [51] give a more up-to-date treatment of the bilinear maps used in cryptographic schemes, along with a detailed discussion of their efficiency.

Our proof of security for the Boneh-Boyen scheme was adapted directly from their paper [18]. For the Waters scheme [109], the proof given here more closely follows an alternate approach suggested by Bellare and Ristenpart [8] although we stress that the same basic ideas were used already by Waters. We remark that is both these aforementioned works [109, 8], the proofs are significantly complicated because they provide a security analysis for identity-based encryption. In contrast, here we faced the easier task of proving security as a signature scheme only.

Part III
Digital Signature Schemes in the Random Oracle Model

Chapter 6
The Random Oracle Model

The signature schemes described in the previous chapters, whether based on the RSA/strong RSA assumptions or bilinear maps, represent essentially the extent of what is currently known regarding *efficient* yet *provably secure* signature schemes. These schemes have some clear disadvantages:

- They are based on a relatively small set of assumptions. Notably, we are lacking efficient constructions based on the hardness of factoring or the discrete logarithm problem, or even the computational Diffie-Hellman assumption in groups without an associated bilinear map.
- Though efficient, the existing schemes are currently viewed as not being efficient *enough* for most practical applications. This is especially true given the availability of various "heuristically secure" schemes (i.e., schemes whose security has not been broken thus far) that are much more efficient than the provably secure schemes we know.

Given what appears to be the *status quo* — that the only schemes having a chance of being widely used are those whose efficiency is truly comparable with currently deployed schemes — we are left with somewhat of a dilemma. On the one hand we would surely prefer to use schemes having *some* concrete evidence for their security; on the other hand the schemes that we can (currently) prove secure are not (yet) efficient enough.

One approach to addressing this dilemma is to introduce new number-theoretic assumptions; these may, in turn, enable constructions of more efficient schemes based on those assumptions, or might allow for proofs of security for existing schemes. (To some extent, this has already happened with the introduction of the strong RSA assumption and — even more so — with the variety of assumptions that have been proposed in the context of groups over which a bilinear map is defined.) This is certainly a reasonable approach. Of course, that there is always the danger that a new problem turns out to be not quite as hard as it appeared on first glance! There is also a more subtle problem that arises when introducing a new assumption expressly to prove an existing scheme secure: at what point does introducing a new assumption become "equivalent" to simply assuming the security of the signature scheme itself? This is not a question that can be answered readily (see [87] for one possible approach), but the point remains that it is preferable to base schemes on well-studied, widely accepted assumptions rather than newer ones.

Another approach to resolving the dilemma has become popular, though the theoretical soundness of the approach has yet to be rigorously justified. This methodology advocates the use of a new model, called the *random oracle model*, in which to prove security of cryptographic schemes.[1] As suggested by the name, the random oracle model is an idealized model that assumes the existence of a public, random function H such that all parties can obtain $H(x)$ (for any desired input value x) only by *interacting with an oracle computing H*; parties *cannot* compute H (for any input) on their own. The oracle (equivalently, H) is assumed to implement a truly random function, something that can be visualized in the following way: the oracle maintains a table of pairs $\{(x_i, y_i)\}$ which is initially empty. When the oracle receives a query x it first checks whether $x = x_i$ for some i; if so, the corresponding y_i is returned. Otherwise, a random string y of the appropriate length is chosen, the answer y is returned, and the oracle stores (x, y) in its table so that the same output value can be returned if the same input is queried again. Queries to the oracle are assumed to be private so that, for example, if an honest party queries the oracle on input x then an adversary does not learn x.

The random oracle model is used to design cryptographic primitives via the following two-step approach:

1. A scheme is designed and proven secure assuming the existence of this random oracle (and possibly also using other number-theoretic or cryptographic assumptions); that is, the scheme is designed in this idealized "random oracle model." We will describe the model in more detail below.
2. In the real world, of course, no random oracle is available. In practice, then, the random oracle is *instantiated* with a cryptographic hash function (such as SHA-1, modified appropriately) that is hoped to be "sufficiently good" at simulating a random oracle.

As we shall see, it is not clear exactly what is meant for a hash function to be "sufficiently good" at simulating a random oracle, nor is it clear that this is a goal that can possibly be achieved in general. Thus, proofs of security for a given scheme in the random oracle model should not be viewed as proofs that apply to any "real-world" instantiation of the scheme. Before discussing in more detail the theoretical drawbacks of the random oracle model, as well as the practical justifications for using it, we will describe by way of example why the random oracle model is so useful in proving security of cryptographic schemes.

---

[1] Indeed, use of the random oracle model is not limited to the case of signature schemes.

6.1 Security Proofs in the Random Oracle Model

At a high level, the random oracle model is used in the following way: given an adversary A breaking some scheme/violating some cryptographic assumption in the random oracle model, A is transformed to an algorithm A' which violates some cryptographic assumption in the standard model *by having A' simulate the random oracle for A*. Our algorithm A' gains two advantages here: (1) A' gets to *see the queries* that A makes to the random oracle; and (2) A' gets to *answer these queries* "any way it likes" (although to provide a good simulation these queries should be answered in a "random-looking" way). There is also a third, perhaps more obvious advantage (already implicit in the random oracle model itself): the value of $H(x)$ for any point x not explicitly queried by A is completely random, at least from the perspective of A.

We give two quick examples of proofs where the above techniques are used in an essential way. We make no attempt to be rigorous here; rather, our goal is to provide the reader with enough intuition so that the later proofs in the random oracle model (given in the chapters that follow) will be easier to digest.

First example. For the first example we consider the following game: a "challenger" generates an RSA key (N, e), chooses random $m \in \{0, 1\}^{\ell}$ and $r \in \mathbb{Z}_N^*$, and sends $(N, e, r^e \bmod N, H(r) \oplus m)$ to A, where $H : \mathbb{Z}_N^* \to \{0, 1\}^{\ell}$ is taken to be a random oracle. We claim that, under the RSA assumption, no PPT adversary A can learn *any information* about m (beyond what it could determine by random guessing) except with negligible probability. To see this, we argue as follows: The *only* way A can learn anything about m is by explicitly querying the random oracle at the point r; this is a consequence of the fact that $H(r)$ is completely random (from the point of view of A) unless A specifically queries H at this point. Letting ε denote the probability that A does indeed query $H(r)$, we have that A learns no information about m except with probability ε.

We now show that ε must be negligible by constructing a PPT algorithm A' which solves the RSA problem with probability exactly ε:

> **Algorithm A'**
> The algorithm is given (N, e, y), and its goal is to compute $y^{1/e} \bmod N$.
>
> - Choose random $r' \leftarrow \{0, 1\}^{\ell}$.
> - Run A on input (N, e, y, r'). Perfectly simulate the random oracle queries of A by returning random answers to each (fresh) query.
> - When A is done, let $x_1, \ldots, x_n$ denote the queries that A made to the random oracle.
> - If there exists an i for which $x_i^e \bmod N \stackrel{?}{=} y$, then output x_i.

The input given to A is distributed identically to the above experiment, and it is trivial for A' to simulate the random oracle for A (since A makes only polynomiallymany queries, A' need only store a polynomial-size table of input/output pairs). Since A' can see the queries made by A to the random oracle, it follows that A' inverts its given input y with probability exactly ε. We conclude that, under the RSA assumption, ε is negligible as claimed.

Second example. We now consider a game in which the challenger generates (N, e) as before, and sends these and a random value $y \in \mathbb{Z}_N^*$ to an adversary A. The goal is for A to compute $y^{1/e} \bmod N$. The "twist" is that A can query the random oracle $H : \mathbb{Z}_N^* \to \mathbb{Z}_N^*$ at any sequence of points $x_1, \ldots, x_\ell \in \mathbb{Z}_N^*$ (we assume without loss of generality that these are distinct), receiving in return the output values $y_1 = H(x_1), \ldots, y_\ell = H(x_\ell)$; the challenger then gives to A the values $y_i^{1/e} \bmod N$ for all i (assume the challenger knows the factorization of N and so can compute these values). We claim that A still cannot compute $y^{1/e} \bmod N$ except with negligible probability. To see this, construct the following adversary A' which computes $y^{1/e} \bmod N$ with the same probability at A, but *without* any additional "help" from the challenger:

> **Algorithm A'**
> The algorithm is given (N, e, y) as input, and its goal is to compute $y^{1/e} \bmod N$.
>
> - Run A on input (N, e, y).
> - For each random oracle query $H(x_i)$ of A:
>   - Choose $r_i \leftarrow \mathbb{Z}_N^*$.
>   - Answer the query using $y_i := r_i^e \bmod N$.
> - Give $r_1, \ldots, r_\ell$ to A. Output whatever value is output by A.

The key point to notice in the above is that A' perfectly simulates the random oracle queries of A (indeed, each answer y_i is uniformly distributed in $\mathbb{Z}_N^*$), but in such a way that A' knows the corresponding inverses $r_i = y_i^{1/e} \bmod N$. Thus, the view of A above is identically distributed to its view in the real experiment, and so A' outputs the correct inverse of y with exactly the same probability with which A outputs the correct inverse in the original game.

6.2 Is the Random Oracle Methodology Sound?

The proofs we have given in the previous two examples are rather simple and clean. It is worth noting, however, that the proofs rely *heavily* on the random oracle model, and cannot be translated to the "real world" when the random oracle H is instantiated with *any* concrete hash function.[2] In the first example, the proof relies on the fact that $H(r)$ is completely random unless r is explicitly queried to the oracle, and on

---

[2] This is not to say that there might not be some other way to prove the same results as in the two examples; the point is only that the *proofs* given earlier do not work.

the ability of A' to see the queries being made by A. Neither of these assumptions hold for any concrete function H which can be evaluated by A on its own. First of all, it is not even clear in this case what it means for A to "explicitly query $H(r)$," since A may be able to evaluate $H(r)$ other than by explicitly making a query to a subroutine that computes H. Given this, it is doubly unclear what it might mean for $H(r)$ to be "completely random" from the point of view of A — note in particular that once A is given a program to compute H, the value $H(r)$ is decidedly no longer random. Finally, once A no longer needs to make explicit queries to compute $H(r)$ there is no way for A' to "intercept" these queries and check for an inverse of y.

The proof of the second example, above, relies on the ability of A' to *answer* the random oracle queries of A in any manner of its choosing. Once the function H is fixed (in particular, once A is given the code to compute H), however, this is of course no longer possible.

The objections raised above are representative of a more general problem with the random oracle model. Recall the two-step approach to designing schemes in this model: first, a scheme is developed and proven secure in the random oracle model; next, the scheme is instantiated in the real world using a hash function which is assumed to provide a "sufficiently good" simulation of a random oracle. Unfortunately, we do not in general know how to prove that *any* concrete hash function is "sufficiently good" at simulating a random oracle; thus, a proof of security in the random oracle model does not actually constitute a proof of security (for the instantiated scheme) in the real world. Worse, we cannot in general even *define* (in a meaningful way) what "simulating a random oracle" means. Because of this, using the random oracle model to "prove" security of a scheme is *qualitatively* different from, e.g., introducing a new cryptographic assumption in order to prove some scheme secure, and a proof of security in the random oracle model is generally regarded as less desirable than a proof of security in the so-called[3] *standard model* which does not allow for random oracles. (The division of the chapters in this book is meant to emphasize this preference for proofs that avoid random oracles.)

It is worth stressing at this point that a random oracle is fundamentally *different* from a pseudorandom function, and in particular a pseudorandom function cannot be used generically to instantiate a random oracle. A pseudorandom function is a *keyed* function which can only be evaluated (in the real world) when the key is known, and is only "random-looking" when the key is *un*known. In contrast, a random oracle is unkeyed and, when instantiated by a hash function in the real world, can be evaluated by anyone; yet it is supposed to remain "random-looking" in some ill-defined sense. Finally, pseudorandom functions can be (meaningfully) defined and realized; in contrast, it is clear that *no* real-world function will have the properties of a random oracle that were relies upon in the proofs above. (As in footnote 2, this is not to say that there does not exist any real-world function for which the previous *claims of security* hold; only that there is no real-world function for which the previous *proofs* will work.)

---

[3] In the use of "standard model" to designate the non-random oracle model, one can already sense the unease with which the random oracle model is regarded.

In light of the above, a major current research question is what, exactly, a proof of security in the random oracle model *does* guarantee vis-a-vis the security of the scheme in the real world. It is fair to say that we are a long way from a good understanding of the answer to this question.

6.2.1 Negative Results

In fact, a number of negative results concerning the use of the random oracle model are known. These negative results generally show a (contrived) scheme which is provably-secure in the random oracle model but demonstrably insecure for *any* concrete instantiation of the random oracle. (In at least one case [89] a stronger result is demonstrated whereby a security goal is shown to be achievable in the random oracle model, but not achievable — by any scheme — in the standard model.) We will give the general flavor of these results, again without being completely formal.

As a warm-up, we first show a weaker result: that for any real-world hash function H^* there exists a signature scheme which is secure in the random oracle model but insecure when instantiated using H. The idea is quite simple: let $H^*(0) = v_0$, where the output-length of H^* is assumed to be linear in the security parameter. Now take any secure signature scheme and modify it so that, when signing a message, the signer first checks whether $H(0) \stackrel{?}{=} v_0$ (where H is the random oracle); if so, the signer outputs its secret key (if not, then the signature is computed normally). In the random oracle model, this modified scheme is still secure: the probability that $H(0) = v_0$ for any value v_0 is negligible, and assuming this is not the case then the signature scheme is unchanged. On the other hand, the scheme is clearly insecure when H is instantiated using H^*.

Along the same general lines, but using a more technically difficult proof, one can show the following, stronger result: there exists a signature scheme which is secure in the random oracle model, but which is insecure when instantiated using any real-world hash function H^*. Take any secure signature scheme and modify it in the following way: when signing a message m the signer first interprets m as the code for a function f_m (if m cannot be interpreted as a program, the signer can take f_m as, say, the identity function). The signer then chooses random values $r_1, \ldots, r_\ell$ for some sufficiently-large (but polynomial) value ℓ, and checks whether $f_m(r_1) \stackrel{?}{=} H(r_1), \ldots, f_m(r_\ell) \stackrel{?}{=} H(r_\ell)$. If so, then the signer outputs its secret key; otherwise, the signature is computed normally.

It is easy to see that the scheme is insecure for any concrete choice of H^*, as an adversary can simply request that the signer sign the code for the program computing H^*. On the other hand, intuitively the scheme is secure in the random oracle model since there is unlikely to *exist* any "short" program m for which f_m agrees with the random oracle on a large fraction of their inputs. (We have glossed over many technical details; the interested reader is referred to [25, 14].) Note that this "disconnect" relies exactly on the fact that a truly random function (as realized by a random oracle) is an *exponential-size* object, while any concrete function instantiating a random oracle must have polynomial size (that is, there must be a polynomial-size program computing this function).

6.3 The Random Oracle Model in Practice

Given the theoretical problems with the random oracle model, as well as the negative results discussed above, are schemes with proofs of security in the random oracle model any better than schemes having no proof of security at all? Although there are those who would disagree, we believe that they are, and offer the following reasons in support of this conviction:

- A proof of security for a given scheme in the random oracle model indicates that the design of the scheme is "sound," in the sense that the only possible weakness in (the real-world instantiation of) the scheme can arise due to a weakness in the hash function used to instantiate the random oracle. Alternately, the only way to "break" the scheme is to "break" the hash function itself (in some way); thus, if the hash function is "good enough" we have some confidence in the security of the scheme itself. Furthermore, if a given instantiation of the scheme *is* successfully cryptanalyzed, we can replace the hash function being used with a "better" one.
- Finally, *there have been no real-world attacks on any "natural" scheme proven secure in the random oracle model* (this is meant to rule out the attacks on the "contrived" schemes such as those discussed in Section 6.2.1). This gives evidence to the usefulness of the random oracle model in designing schemes in practice.
- In light of the above, a proof of security in the random oracle model is preferable to no proof at all. Of course, this assumes that these are the only options; i.e., that schemes with proofs of security in the standard model will not be used due to their inefficiency. (We do not argue that choosing efficiency at the expense of security is the right decision to make, only that it seems to be the decision made most frequently in practice. This also assumes that the schemes proven secure in the random oracle model are indeed significantly more efficient than known schemes with proofs in the standard model.)

Nevertheless, the above ultimately represent only intuitive speculation as to the usefulness of the random oracle model rather than rigorous proof, and we emphasize our opinion that proofs of security in the standard model are preferable to proofs in the random oracle model. Understanding exactly what such proofs guarantee in the real world remains, in our mind, one of the most important research questions facing cryptographers today.

6.4 Further Reading

See [72, Chapter 13] for extensive additional discussion of the pros and cons of
the random oracle model, examples of its application to security proofs, and some
comments regarding implementation of random oracles in the real world.

The use of random oracle model to facilitate proofs was first suggested by Fiat
and Shamir [47]. The model was formalized, advocated, and popularized by the
seminal work of Bellare and Rogaway [9], after which the use of the random oracle
model to prove security of cryptographic constructions truly began to take off.

For negative results concerning the random oracle model, see [25, 14].

Chapter 7
Full-Domain Hash (and Related) Signature Schemes

An important class of signature schemes proven secure in the random oracle model is given by the *full-domain hash* (FDH) signature scheme and its variants. In addition to being simple and natural, as well as quite efficient, constructions in this family are also the basis for standardizes signature schemes that are widely used.

We begin by describing the basic FDH signature scheme, which can be instantiated with any doubly enhanced[1] trapdoor permutation. (The same ideas can also be instantiated using groups over which a bilinear map is defined.) We then focus on techniques for improving the *tightness* of the security reduction, i.e., the "gap" between the probability with which an adversary forges a signature and that with which the security reduction solves the underlying hard problem on which the scheme is based. (This will be explained more concretely after our treatment of FDH.) Interestingly, these techniques all rely on the stronger assumption of (doubly enhanced) *clawfree* trapdoor permutations (though they are all described here using RSA as a special case).

We first show that a tighter analysis of FDH is possible. Even this reduction has some slack, however, and so we turn to other schemes that offer optimally right reductions. The first of these is a randomized variant of FDH where a random "salt" is chosen (and included with the signature) each time a message is signed. We then show how the salt can be avoided while maintaining the tight reduction.

7.1 The Full-Domain Hash (FDH) Signature Scheme

Full-domain hash (FDH) gives what is perhaps the most intuitively appealing approach to constructing digital signatures, and can be viewed as a secure realization of the original ideas of Diffie and Hellman. At a high level, the public key in FDH consists of (a description of) a trapdoor permutation f; the secret key is (the de-

---

[1] In the literature, FDH is usually described as being based on trapdoor permutations. As we will see in the proof of Theorem 7.1, however, *doubly enhanced* trapdoor permutations are needed.

scription of) the inverse f^{-1}. Letting H be a hash function (that will be modeled as a random oracle) mapping messages to the domain of f, the signature on a message m is simple $f^{-1}(H(m))$. Verification of a candidate signature σ can be done by simply checking whether $f(\sigma) \overset{?}{=} H(m)$. This is formalized as Construction 7.1, where the main difference is that we are more precise regarding our use of H..

Construction 7.1: The FDH signature scheme

Let $\Pi = (\mathsf{Gen}, \mathsf{Samp}, f, f^{-1})$ be a trapdoor permutation family, and let $H : \{0,1\}^k \to \{0,1\}^k$ be a hash function.

Key generation: Compute $(I, \mathsf{td}) \leftarrow \mathsf{Gen}(1^k)$. The public key is I and the secret key is td.

Signature generation: To compute the signature on a message m, compute $r := H(m)$ followed by $y := \mathsf{Samp}(I; r)$. Then output the signature $f_{\mathsf{td}}^{-1}(y)$.

Signature verification: To verify a signature σ on a message m, compute $r := H(m)$ followed by $y := \mathsf{Samp}(I; r)$. Then output 1 iff $f(\sigma) \overset{?}{=} y$.

Theorem 7.1. *If Π is a doubly enhanced trapdoor permutation, and H is modeled as a random oracle, then Construction 7.1 is strongly unforgeable under an adaptive chosen-message attack.*

Proof. Messages in FDH have unique signatures, so unforgeability implies strong unforgeability. Given a PPT adversary A attacking the FDH signature scheme, making q_H hash queries (i.e., queries to the random oracle computing H) and forging a signature with probability ε, we construct a PPT algorithm A' inverting Π (in the sense described in Definition 2.5) with probability at least ε/q_H. Since q_H must be polynomial, we conclude that ε must be negligible.

Because Π is doubly enhanced, there exists (cf. Definition 2.5) an algorithm Samp' that takes as input I and outputs a tuple (x, y, r) such that (1) $f_I(x) = y$ and $y = \mathsf{Samp}(I; r)$, and (2) the distribution on r is statistically close to uniform. For simplicity we will assume that the distribution on r is uniform. We also make three assumptions, without loss of generality, regarding the behavior of A:

1. A never repeats a query to the random oracle.
2. Before making a signature query on a message m, adversary A first makes the hash query $H(m)$.
3. Before outputting a forgery (m, σ), adversary A first makes the hash query $H(m)$.

Our algorithm A' proceeds as follows:

> **Algorithm A':**
> The algorithm is given (I, y, r) as input, with $y = \mathsf{Samp}(I; r)$.
> Its goal is to output x such that $f_I(x) = y$.

- Choose $j \leftarrow \{1, \ldots, q_H\}$.
- Run A on the public key I. Store triples $(\cdot, \cdot, \cdot)$ in a table, initially empty. An entry (m_i, x_i, r_i) indicates that A' has set $H(m_i) = r_i$, and $f_I(x_i) = \mathsf{Samp}(I; r_i)$ (in particular, x_i is a valid signature on m_i).
- When A makes its ith random oracle query $H(m_i)$, answer it as follows:
  - If $i = j$, return r.
  - Otherwise, compute $(x_i, y_i, r_i) \leftarrow \mathsf{Samp}'(I)$, return r_i as the answer to the query, and store (m_i, x_i, r_i) in the table.

  When A requests a signature on a message m, let i be such that $m = m_i$ and answer this query as follows:
  - If $i \neq j$, then there is an entry (m_i, x_i, r_i) in the table. Return x_i.
  - If $i = j$ then abort.
- If A outputs a forgery (m, σ) with $m = m_j$, return σ.

We may easily observe that as long as A' does not abort, the simulation it provides for A is perfect. Specifically:

1. The public key has the correct distribution.
2. A's jth query to the random oracle is answered with the random string r. By definition of Samp', all of A's other queries to the random oracle are also answered with a random string.
3. Signing queries (assuming A' does not abort) are also answered correctly; this, again, follows from the properties of Samp'.

Moreover, if A outputs a forgery (m, σ) with $m = m_j$ then

$$f_I(\sigma) = \mathsf{Samp}(I; H(m_j)) = \mathsf{Samp}(I, r) = y,$$

and so A' correctly solves its given instance. The theorem follows from the fact that the guess j made by A' (representing a guess of the hash-query index for which A will produce its forgery) is correct with probability $1/q_H$.

7.1.1 An Instantiation Using Bilinear Maps

For completeness, we show here how the FDH approach can be instantiated using bilinear maps (even though these do not quite give a trapdoor permutation family). The scheme described here was introduced by Boneh, Lynn, and Shacham and is referred to as the *BLS signature scheme*. For simplicity in the description that follows, we assume a random oracle mapping directly to the appropriate group; in practice, one has to be careful to instantiate this random oracle properly.

Construction 7.2: The BLS signature scheme

Let $\mathcal{G}$ be as in Chapter 5.

Key generation: Compute params $= (\mathbb{G}, \mathbb{G}_T, q, g, \hat{e}) \leftarrow \mathcal{G}(1^k)$. Choose $x \leftarrow \mathbb{Z}_q$ and set $y := g^x$. The public key is (params, y) and the secret key is x.

Let $H : \{0,1\}^k \rightarrow \mathbb{G}$ be a hash function.

Signature generation: To compute the signature on a message m, compute $h := H(m)$ and output the signature h^x.

Signature verification: To verify a signature σ on a message m, check whether $\hat{e}(\sigma, g) \stackrel{?}{=} \hat{e}(H(m), y)$.

Theorem 7.2. *If the CDH problem is hard relative to $\mathcal{G}$, and H is modeled as a random oracle, then Construction 7.2 is strongly unforgeable under an adaptive chosen-message attack.*

Proof. The proof here is substantially identical to the proof of Theorem 7.1, and so we only sketch the analysis. Given a PPT adversary A attacking the BLS signature scheme, making q_H hash queries (i.e., queries to the random oracle computing H) and forging a signature with probability ε, we construct a PPT algorithm A' solving the CDH problem with probability at least ε/q_H. We make the same assumptions on the behavior of A as in the proof of Theorem 7.1.

Our algorithm A' proceeds as follows:

> **Algorithm A':**
> The algorithm is given $(\mathbb{G}, \mathbb{G}_T, q, g, \hat{e}, y, h)$ as input, and its goal is to output $\sigma \in \mathbb{G}$ with $\hat{e}(\sigma, g) = \hat{e}(h, y)$.
>
> - Choose $j \leftarrow \{1, \ldots, q_H\}$.
> - Run A on the public key (params, y). Store triples $(\cdot, \cdot, \cdot)$ in a table, initially empty. An entry (m_i, x_i, h_i) indicates that A' has set $H(m_i) = h_i$, and $h_i = g^{x_i}$.
> - When A makes its ith random oracle query $H(m_i)$, answer it as follows:
>   - If $i = j$, return h.
>   - Otherwise, choose $x_i \leftarrow \mathbb{Z}_q$, set $h_i := g^{x_i}$, return h_i as the answer to the query, and store (m_i, x_i, h_i) in the table.
>   When A requests a signature on a message m, let i be such that $m = m_i$ and answer this query as follows:
>   - If $i \neq j$, then there is an entry (m_i, x_i, h_i) in the table. Return y^{x_i}.
>   - If $i = j$ then abort.
> - If A outputs a forgery (m, σ) with $m = m_j$, return σ.

It is not difficult to verify that, unless A' aborts, the simulation provided for A is perfect and A' correctly solves its given CDH instance if A forges a signatures. The theorem follows.

7.2 An Improved Security Reduction for FDH

The proofs of security we have given for the FDH and BLS signature schemes provide the following guarantee: if there exists a PPT adversary A making q_H hash queries who can "break" the signature scheme with probability ε, then there exists a PPT algorithm A' that can solve the underlying hard problem with probability ε/q_H. (In general, the running time of the reduction A' as a function of the running time of A is also an important concern. In the proofs we have given, however, the running time of A' was within a relatively small multiple of the running time of A.) Turning this around, this means that if we believe that some relevant computational problem is hard to solve in time t with probability better than ε' (for some concrete setting of the security parameter k), then using this computational problem in one of these constructions gives a signature scheme that cannot be broken with probability better than $q_H \varepsilon'$. Since q_H, in practice, corresponds to the number of hash function evaluations made by A, this can lead to a significant loss in security! For example, one may set parameters so that $\varepsilon' \approx 2^{-100}$. But then if $q_H \approx 2^{80}$ (a large, but not unachievable, value), the signature scheme only has security $\varepsilon \approx 2^{-20}$ (a not-very-conservative choice). In theory, the schemes are still "secure" since we may take the security parameter as high as we like (and are assured that, asymptotically, the advantage of any adversary will decay quickly); in practice, however, we see that the *tightness* of a security reduction is of extreme importance.

Interestingly, the security reduction for the FDH signature scheme can be improved *when a clawfree trapdoor permutation family* is used. (A similar idea can be applied directly to the BLS signature scheme, using the observation — recorded in Section 2.3.3 — that groups in which the discrete logarithm problem is hard imply clawfree permutations.) It can be shown that if there exists a PPT adversary A making q_S *signing* queries who can "break" the signature scheme with probability ε, then there exists a PPT algorithm A' that can solve the underlying hard problem with probability $O(\varepsilon/q_S)$. Since $q_S \ll q_H$, this represents a significant improvement.

We illustrate the idea here using the specific example of RSA as a (doubly enhanced) clawfree trapdoor permutation (cf. Section 2.2.2). Here, the public key is (N, e) and the signature on a message m is $H(m)^{1/e} \bmod N$. Examining the reduction A' presented in the proof of Theorem 7.1, and assuming for simplicity that the hash function H maps directly onto $\mathbb{Z}_N^*$, we see that A' will guess the index i of the hash query that results in a forgery; A' then sets the output of the ith hash query $H(m_i)$ to be y (the value to be inverted). The output of all other hash queries $H(m_j)$ is set to $x_j^e \bmod N$ for random $x_j \in \mathbb{Z}_N^*$. In this way, A' can answer all signing queries except those for message m_i; furthermore, if adversary A outputs a forged signature on m_i, then A' obtains the desired inverse. The loss of security arises from having to guess the index i from among all q_H hash queries of A.

We can improve the reduction by having A' be a bit more clever in how it responds to hash queries of A. Specifically, each hash query will now be answered in one of two ways. With probability γ, a hash query is answered as before (i.e., $H(m) = x^e \bmod N$ for a random x); let us refer to any message whose hash is computed the way as a *type-0* message. With probability $1 - \gamma$, however, a hash query is answered by choosing random $x \in \mathbb{Z}_N^*$ and setting the answer to be $yx^e \bmod N$; we refer to any message whose hash is computed this way as a *type-1* message. Note that A' can compute a valid signature, exactly as before, on any type-0 message; a forged signature on a type-0 message, though, is useless to A'. In exactly the opposite way, A' is unable to compute a valid signature on any type-1 message, but a forgery on any such message allows A' to compute the desired inverse as we now describe. Say A forges a valid signature σ on a type-1 message m whose hash was computed as $H(m) = yx^e \bmod N$. Then

$$(\sigma/x)^e = \sigma^e/x^e = H(m)/x^e = y \bmod N,$$

and so $\sigma/x \bmod N$ is the correct answer. For completeness, we give the reduction here:

Algorithm A':
The algorithm is given (N, e, y) as input, and its goal is to output x such that $x^e = y \bmod N$.

- Run A on the public key (N, e). Store triples $(\cdot, \cdot, \cdot)$ in a table, initially empty. An entry (b_i, m_i, x_i) indicates that m_i is a type-b_i message, answered using randomness x_i (see below).
- When A makes its ith random oracle query $H(m_i)$, choose random $x_i \leftarrow \mathbb{Z}_N^*$ and answer the query as follows:
  - With probability γ, return $x_i^e \bmod N$ as the answer to the query, and store $(0, m_i, x_i)$ in the table.
  - With probability $1 - \gamma$, set $b_i := 2$, return $yx_i^e \bmod N$ as the answer to the query, and store $(1, m_i, x_i)$ in the table.
  When A requests a signature on a message m, let i be such that $m = m_i$ and let (b_i, m_i, x_i) be the corresponding entry in the table. Answer this query as follows:
  - If $b_i = 0$, then return x_i. If $b_i = 1$, then abort.
- If A outputs a forgery (m, σ), let i be such that $m = m_i$ and let (b_i, m_i, x_i) be the corresponding entry in the table. If $b_i = 0$ abort. Otherwise output $\sigma/x_i \bmod N$.

A' aborts if A ever asks for a signature on any type-1 message, or if A' outputs a forgery on a type-0 message. The probability with which A' does not abort is thus

$$p \stackrel{\text{def}}{=} \gamma^{q_S} \cdot (1 - \gamma),$$

which is maximized at $p = O(1/q_S)$ by setting $\gamma = q_S/(q_S + 1)$. Putting everything together, if A outputs a valid forgery with probability ε then A' solves its given RSA instance with probability $O(\varepsilon/q_S)$ as claimed.

For the generalization of the above to the case of arbitrary clawfree trapdoor permutations, see the references at the end of this chapter.

7.3 Probabilistic FDH

The result described in the previous section still leaves a gap. Here, we show a probabilistic variant of FDH (called PFDH) that achieves a *tight* security reduction; namely, this scheme has a proof of security showing that an adversary breaking the scheme with probability ε yields an algorithm solving the underlying computational problem with probability $\varepsilon' \approx \varepsilon$. Here, too, the stronger assumption of clawfree trapdoor permutations is needed; Construction 7.3 gives an instantiation of the approach using RSA. (We discuss how to set κ as part of the security proof below.)

Construction 7.3: The PFDH-RSA signature scheme

Let GenRSA be as in Chapter 4.

Key generation: Compute $(N, e, d) \leftarrow \mathsf{GenRSA}(1^k)$. The public key is (N, e) and the secret key is d.

Let $H : \{0, 1\}^k \to \mathbb{Z}_N^*$ be a hash function.

Signature generation: To compute the signature on a message m, choose random $r \leftarrow \{0, 1\}^\kappa$, compute $y := H(r \| m)$, and output the signature $(r, y^d \bmod N)$.

Signature verification: To verify a signature (r, σ) on a message m, check whether $\sigma^e \stackrel{?}{=} H(r \| m)$.

The key novelty here is that messages now have *multiple* valid signatures, and the reduction A' can use this to its advantage. Let (N, e, y) be the RSA instance given to A'. In the reduction that follows, A' gives the public key (N, e) to A and (essentially) computes in advance a list of q_S signatures for each possible message m. (To avoid exponential running time, A' determines the list of signatures for a message m only after m is used in a hash query of A.) That is, for each message m the reduction A' chooses random $r_1, \ldots, r_{q_S}$, sets $H(m \| r_i) = x_i^e \bmod N$ for all i (with the $\{x_i\}$ chosen randomly and independently, except that $x_i = x_j$ if $r_i = r_j$), and fixes the ith signature on m to be (r_i, x_i). The $r_1, \ldots, r_{q_S}$ are stored in a list L_m. Using the terminology from the previous section, any hash query $H(m \| r)$ with $r \in L_m$ is thus answered as a type-0 query. A hash query $H(m\|r)$ with $r \notin L_m$ is answered as a type-1 query; that is, A' sets the answer to be $yx^e \bmod N$ for random $x \in \mathbb{Z}_N^*$. We see that A' can answer all signing queries of A without any possibility of abort. Moreover, if A outputs a forgery (r, σ) on a message m, then A' can output the desired solution to its RSA instance iff $r \notin L_m$; we then only need to argue that this happens with sufficiently high probability.

We now describe the reduction formally:

> **Algorithm A':**
> The algorithm is given (N, e, y) as input, and its goal is to output x such that $x^e = y \bmod N$.
>
> - Run A on the public key (N, e). Store triples $(\cdot, \cdot, \cdot)$ in a table that is initially empty.
> - When A makes oracle query $H(m\|r)$ do:
>   1. If there is an entry $(m\|r, x, z)$ in the table, return z.
>   2. If list L_m already exists, go to the next step. Otherwise, choose q_S values $r_{m,1}, \ldots, r_{m,q_S} \leftarrow \{0, 1\}^\kappa$ and store them in a list L_m.
>   3. If $r \in L_m$ then let i be such that $r = r_{m,i}$. Choose random $x_{m,i} \in \mathbb{Z}_N^*$ and return the answer $z = x_{m,i}^e \bmod N$. Store $(m\|r, x_{m,i}, z)$ in the table.
>   4. If $r \notin L_m$, choose random $x \in \mathbb{Z}_N^*$ and return the answer $z = yx^e \bmod N$. Store $(m\|r, x, z)$ in the table.
>
>   When A requests some message m to be signed for the ith time, let $r_{m,i}$ be the ith value in L_m and compute $z = H(m\|r_{m,i})$ as above if the hash query $H(m\|r_{m,i})$ has not yet been asked. Let $(m\|r_{m,i}, x_{m,i}, z)$ be the corresponding entry in the table. Output the signature $(r_{m,i}, x_{m,i})$.
> - If A outputs a forgery $(m, (r, \sigma))$ and $r \in L_m$ then abort. Otherwise, let $(m\|r, x, z)$ be the corresponding entry of the table, and output $\sigma/x \bmod N$.

As noted above, A' provides a perfect simulation for A, and so if A outputs a forgery with probability ε in a real attack on the signature scheme then it outputs a forgery with the same probability in the above experiment. Conditioned on A's outputting a valid forgery (r, σ) on a message m, the reduction A' succeeds iff $r \notin L_m$. Since each element in L_m is chosen uniformly, this occurs with probability exactly $p = (1 - 2^{-\kappa})^{q_S}$. Setting $kappa = \log q_S$ (and assuming $q_S \geq 2$), we have $p \geq 1/4$. Overall, then, A' outputs the correct solution to its given RSA instance with probability at least $\varepsilon/4$, and the reduction is therefore tight.

We remark that the same argument, applied to the same reduction, shows that PFDH-RSA is in fact *strongly* unforgeable.

7.4 A Simpler Variant with a Tight Reduction

The PFDH scheme in the previous section uses a random salt, included as part of the signature, each time a message is signed. Here we show a simple technique that maintains the tight reduction while avoiding the salt altogether. As in the previous two sections, the basic idea here can be instantiated using any clawfree trapdoor permutation but we illustrate the idea using the specific example of RSA.

Construction 7.4: A (deterministic) variant of FDH

Let GenRSA be as in Chapter 4.

Key generation: Compute $(N, e, d) \leftarrow \text{GenRSA}(1^k)$ and then choose a random element $y \in \mathbb{Z}_N^*$. The public key is (N, e, y) and the secret key is d.

Let $H : \{0,1\}^k \to \mathbb{Z}_N^*$ be a hash function.

Signature generation: To compute the signature on a message m, compute $z := H(m)$ and choose a random bit $b \in \{0,1\}$. Then:

- If $b = 0$ output $z^d \bmod N$.
- If $b = 1$ output $(yz)^d \bmod N$.

If a given message is ever signed more than once, the same signature is released each time (see text).

Signature verification: A signature σ on a message m is accepted as valid if either

$$\sigma^e \stackrel{?}{=} H(m) \bmod N \quad \text{or} \quad \sigma^e \stackrel{?}{=} y \cdot H(m) \bmod N.$$

In order to obtain a tight security proof, it is essential that only one signature is released for any given message. (Interestingly, however, the scheme remains secure [with a non-tight reduction] even if both possible signatures are released.) In practice this could be done, without maintaining state, by including a key K for a pseudorandom function as part of the secret key and then determining the bit b for a given message m as $b := F_K(m)$.

The reduction here is actually quite simple. Let us call a signature σ on a message m a *type-0* signature if $\sigma^e = H(m) \bmod N$, and a *type-1* signature if $\sigma^e = yH(m) \bmod N$. For each message m, the reduction A' chooses a random bit b in advance and then sets $H(m)$ in such a way that it can compute a type-b signature for m. If the adversary outputs a forgery, then with probability $1/2$ the forged signature will have the opposite type than the signature that is known by A'. In that case, A' easily computes a eth root of the value y included as part of the public key.

(In fact, the scheme is even strongly unforgeable.) We present the reduction here in its entirety:

Algorithm A':
The algorithm is given (N, e, y) as input, and its goal is to output x such that $x^e = y \bmod N$.

- Run A on the public key (N, e, y). Store tuples $(\cdot, \cdot, \cdot, \cdot)$ in a table that is initially empty.
- When A makes oracle query $H(m)$, choose a random bit $b \in \{0, 1\}$ and a random $x \in \mathbb{Z}_N^*$ and then do:
  1. If $b = 0$, return $z = x^e \bmod N$ as the answer to the query.
  2. If $b = 1$, return $z = x^e/y \bmod N$ as the answer to the query.
  Store (b, m, x, z) in the table.
  When A requests the signature on some message m, find the entry (b, m, x, z) in the table and output x.
- If A outputs a forgery (m, σ), find the entry (b, m, x, z) in the table and then:
  1. If $b = 0$ and $\sigma^e = yz$, then output $\sigma/x \bmod N$.
  2. If $b = 1$ and $\sigma^e = z$, then output $x/\sigma \bmod N$.

If A forges a signature with probability ε, then A' solves its given RSA instance with probability $\varepsilon/2$.

7.5 Further Reading

The FDH signature scheme, inspired by existing signature standards, was presented and analyzed by Bellare and Rogaway [9]. The reduction in Section 7.1 is based on their work. The improved analysis in Section 7.2 is due to Coron [28], who later gave evidence [29] that this reduction is the best possible (for FDH). The idea of using randomized hashing to obtain a tight security reduction is due to Bellare and Rogaway [10], who introduced a scheme called PSS; the PFDH scheme shown here was introduced by Coron [29] as a simplified abstraction of PSS. In the same work, Coron showed how to improve the reduction given in [10] so that a shorter random salt could be used; we have used Coron's tighter analysis (as applied to PFDH) in Section 7.3. PSS-RSA has since been standardized [95]. The FDH variant presented in Section 7.4 adapts ideas of Katz and Wang [73].

Coron's analysis in each of the works just cited was specific to an RSA-based instantiation, and Dodis and Reyzin [41] subsequently showed that the analysis in each case could be extended to any trapdoor clawfree permutation.

The BLS signature scheme appears in [19]. Techniques for hashing onto the appropriate group are discussed in that paper as well as more recent work of Coron and Icart [30].

Factoring-based instantiations of FDH-like schemes, resulting in what are among the most efficient signatures known, are analyzed thoroughly by Bernstein [13].

Chapter 8
Signature Schemes from Identification Schemes

There are currently two main techniques for constructing signature schemes in the random oracle model. The first technique uses the "full-domain hash" approach, and several schemes designed using this approach were introduced in the previous chapter. Here we cover the second central method, in which signature schemes are derived from so-called *identification schemes*. We note up front that there is a rich literature studying identification schemes in their own right; however, we limit ourselves to a discussion of only those aspects that are most directly relevant to the construction of signature schemes.

The chapter is organized as follows. We first define identification schemes as well as the notion of *passive security* for such schemes; although this definition of security is relatively weak as far as identification schemes are concerned, the definition suffices for our purposes. We then describe and prove secure the *Fiat-Shamir transform*, which provides a general method for converting (a certain class of) passively secure identification schemes to signature schemes in the random oracle model. We also discuss a simple transformation of certain identification schemes to KMA-secure signature schemes *without* relying on the random oracle model. Armed with these result, we then devote several sections to identification schemes *per se*: specifically, we show several identification schemes, based on a variety of number-theoretic assumptions, to which the Fiat-Shamir transform can be applied.

We remark that many of the ideas explored in this chapter have proven extremely useful in other areas of cryptography, most notably in the context of *zero-knowledge proofs*. Although we do not describe any of these additional applications, we hope to whet the reader's appetite and thereby motivate the reader to explore these applications; we also expect that the material covered here will provide a firm foundation with which to better understand these more advanced topics.

J. Katz, *Digital Signatures*, DOI 10.1007/978-0-387-27712-7_8,
© Springer Science+Business Media, LLC 2010

8.1 Identification Schemes

Consider a scenario in which a party $\mathscr{P}$ (or *prover*) wants to convince another party $\mathscr{V}$ (or *verifier*) that $\mathscr{P}$ is indeed who he claims to be. One can imagine this arising in many ways: for example, perhaps $\mathscr{P}$ and $\mathscr{V}$ have never met before, or perhaps $\mathscr{P}$ and $\mathscr{V}$ are communicating over a network (and not face-to-face) and $\mathscr{V}$ wants to ensure that he is not communicating with $\mathscr{P}$ rather than an imposter. For the problem to be meaningful there must clearly be some information distinguishing $\mathscr{P}$ from everyone else; otherwise, anyone could easily impersonate $\mathscr{P}$. In a setting in which $\mathscr{P}$ wants to be able to convince *multiple* potential verifiers (rather than just a single, fixed verifier), the natural approach is to assume that $\mathscr{P}$ establishes a public key *pk* which is known to any potential verifier; then, using an associated secret key *sk* generated along with this public key, $\mathscr{P}$ can run an instance of an *identification scheme* to convince $\mathscr{V}$ that he is the party identified with *pk*.

Before defining any notion of security for this setting, let us first give a purely syntactic definition of an identification scheme in order to fix some notation. For two interactive protocols A and B, we let $b \leftarrow \langle A(x), B(y) \rangle$ denote an execution of A (holding input x) with B (holding input y), with b denoted the final output of B at the conclusion of the protocol.

Definition 8.1. An **identification scheme** consists of three probabilistic, polynomial-time algorithms $(\mathsf{Gen}, \mathscr{P}, \mathscr{V})$ such that:

- The randomized **key generation algorithm** Gen takes as input the security parameter k (in unary). It outputs a pair of keys (pk, sk), where pk is called the **public key** and sk is called the **secret key**. We assume the security parameter k is implicit in both pk and sk.
- $\mathscr{P}$ and $\mathscr{V}$ are interactive protocols. The **prover algorithm** $\mathscr{P}$ takes as input a secret key sk and the **verification algorithm** $\mathscr{V}$ takes as input a public key pk. At the conclusion of the protocol, $\mathscr{V}$ outputs a bit b with $b = 1$ signifying "accept" and $b = 0$ signifying "reject."

We require that for all k and all (pk, sk) output by $\mathsf{Gen}(1^k)$ we have:

$$\Pr[\langle \mathscr{P}(sk), \mathscr{V}(pk) \rangle = 1] = 1.$$

Let us explore next what kind of security an identification scheme should achieve. For an identification scheme to be useful, a minimal level of security requires that an adversary A who knows the public key pk of some honest prover $\mathscr{P}$ should be unable to falsely impersonate $\mathscr{P}$ to a verifier. A bit more formally, this would mean that for any efficient adversary A the following should be negligible:

$$\Pr\left[(pk, sk) \leftarrow \mathsf{Gen}(1^k) : \langle A(pk), \mathscr{V}(pk) \rangle = 1 \right]. \tag{8.1}$$

This notion of security is easy to achieve: simply generate the public and secret keys by choosing a random x and setting $sk = x$ and $pk = f(x)$, where f is a one-way function. During an execution of the identification protocol, the prover simply

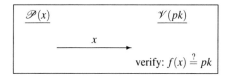

Fig. 8.1 A simple identification protocol.

sends x and the verifier checks whether $pk \stackrel{?}{=} f(x)$ (see Figure 8.1). It is easy to see that this both satisfies the functional definition stated earlier as well as the notion of security expressed in Equation (8.1).

What is wrong with this simple protocol? Well, it clearly does not protect against an *eavesdropping adversary* who monitors even a single execution of the protocol between $\mathscr{P}$ and $\mathscr{V}$. In particular, $\mathscr{P}$ sends its entire secret key in the clear; hence, any adversary who eavesdrops on one execution of the protocol obtains the secret key and can impersonate $\mathscr{P}$ from that point on.

An identification scheme is *secure against passive attacks* if, informally, it remains difficult for an adversary to impersonate $\mathscr{P}$ even after eavesdropping on arbitrarily many executions of the protocol between $\mathscr{P}$ and an honest verifier $\mathscr{V}$. To give a formal definition, we introduce an oracle $\mathsf{Trans}_{sk,pk}(\cdot)$ which, on empty input, returns a transcript (i.e., all messages sent and received) of an honest execution $\langle \mathscr{P}(sk), \mathscr{V}(pk) \rangle$ of the identification scheme. We can thus model each eavesdropping attempt of an adversary by a query to this oracle. Note that if $\mathscr{P}, \mathscr{V}$ are randomized, then Trans is randomized as well and so returns a (possibly) different transcript every time it is invoked. We point out also that Trans *only* returns those messages that would be available to an eavesdropper; in particular, the internal states of the parties (and specifically their random coins) are not included in the information returned by Trans. Finally, we stress that this oracle is modeling eavesdropping attacks on honest executions of the protocol (i.e., executions of $\mathscr{P}$ with an *honest* verifier $\mathscr{V}$); see further discussion below.

With this in place, we now give the formal definition.

Definition 8.2. An identification scheme $(\mathsf{Gen}, \mathscr{P}, \mathscr{V})$ is **secure against a passive attack**, or **passively secure**, if the following is negligible for all PPT adversaries $A = (A_1, A_2)$:

$$\Pr\left[\begin{array}{l}(pk, sk) \leftarrow \mathsf{Gen}(1^k) \\ s \leftarrow A_1^{\mathsf{Trans}_{sk,pk}(\cdot)}(pk)\end{array} : \langle A_2(s), \mathscr{V}(pk) \rangle = 1\right].$$

In the above definition we imagine the adversary carrying out an attack in two "stages": first the adversary eavesdrops on the protocol multiple times (this is reflected by giving A_1 access to the Trans oracle), and it eventually outputs some state information s; the adversary, using s, then tries to impersonate $\mathscr{P}$. This definition is equivalent to the seemingly stronger definition in which we allow A_2 to query the Trans oracle even *while* it is actively trying to impersonate $\mathscr{P}$ (so, for example, if

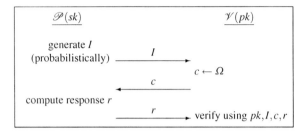

Fig. 8.2 A canonical identification scheme (see text).

the identification scheme is a three-round protocol then A_2 might decide to query the Trans oracle between the second and third rounds). To see that security in this case is implied by Definition 8.2, let A be an adversary strategy (there is clearly no longer any point in separating the adversary into two stages) which makes queries to Trans both before and during its interaction with $\mathcal{V}$, and consider the following strategy $A' = (A'_1, A'_2)$ which achieves identical success probability but does *not* query the Trans oracle during its second stage. Let $q = q(k)$ be a (polynomial) upper bound on the number of times A queries the Trans oracle. Then A'_1 makes q queries to Trans to obtain a sequence of transcripts $t_1, \ldots, t_q$. The output of A'_1 is $s = t_1, \ldots, t_q$. In the second stage, $A'_2(s)$ simply runs A (forwarding messages to/from $\mathcal{V}$ in the obvious way); whenever A makes its ith query to Trans, this is answered by A'_2 using the transcript t_i it has already obtained. Clearly A'_2 makes no queries to Trans, yet A' succeeds in impersonating $\mathcal{P}$ with exactly the same probability as A does.

We will freely use either formulation of Definition 8.2 in what follows.

Before moving on to discuss the connection between identification schemes and signature schemes in the following section, we do want to remark that passive security is indeed a rather weak notion of security: for one, it does not protect against adversaries who pose as verifiers (and thus interact with $\mathcal{P}$ in executions of the identification scheme) and can behave dishonestly as they do so, and then subsequently try to impersonate $\mathcal{P}$ to some other (honest) verifier. Security against such attacks, termed *active security*, as well as security in other, stronger attack models, has received a lot of attention in the literature (see the end of this chapter for some pointers) and such considerations lead to a number of fascinating research directions. Since these stronger definitions of security are irrelevant for our purposes, though, we do not provide any further details here.

Throughout this chapter, we assume identification schemes for which the following hold (see Figure 8.2):

- The identification protocol itself takes three rounds: i.e., an execution of the protocol consists of an initial message I sent by $\mathcal{P}$, a "challenge" c sent by $\mathcal{V}$, and a final response r sent by $\mathcal{P}$.
- We assume the challenge c is chosen uniformly from some set Ω. (Technically, $\Omega = \{\Omega_k\}$ depends on the security parameter k. In general, it may also depend on the public key pk.) This implies that anyone given the transcript of an exe-

cution of the protocol (along with $\mathscr{P}$'s public key pk), can efficiently determine whether $\mathscr{V}$ would have accepted following that execution. By way of terminology, we say that (pk,I,c,r) is an *accepting transcript* if $\mathscr{V}$ would accept an execution of the protocol resulting in this transcript. (When pk is understood from the context, we simply refer to (I,c,r) as an accepting transcript.)

- We assume the fist message of the protocol is "non-degenerate" in the following sense: for any secret key sk and any fixed message $\hat{I}$, the probability that $\mathscr{P}(sk)$ outputs $I = \hat{I}$ as the first message is negligible. (This implies, in particular, that the probability that some first message I repeats in polynomially many executions of the protocol is negligible.) Note that this can easily be achieved for any 3-round identification scheme by simply having $\mathscr{P}$ send an additional k-bit random string (which is ignored by $\mathscr{V}$) as part of its first message.

We refer to identification schemes satisfying the above as *canonical*.

8.2 From Identification Schemes to Signatures

We begin by describing the *Fiat-Shamir transform*, an extremely general technique for converting passively-secure (canonical) identification schemes to signature schemes in the random oracle model. We then isolate two criteria that are sufficient (though not necessary) to assure passive security of an identification scheme. (Although these criteria are not required to apply the Fiat-Shamir transform, relying on these criteria simplifies matters.) We conclude this section by showing how identification schemes satisfying slightly stronger criteria (satisfied by all the identification schemes that will be presented in the following section) can be converted to KMA-secure signature schemes without invoking the random oracle model.

8.2.1 The Fiat-Shamir Transform

The basic idea behind the Fiat-Shamir transform is to have the prover run an instance of the identification protocol *by itself*, generating the challenge c by applying a hash function H to the first message I and then computing an appropriate response r. (Compare Figure 8.3 to Figure 8.2.) If H is modeled as a random oracle, then the challenge c generated by H is "truly random" and so it will be just as difficult for an adversary (who does not know sk) to find an accepting transcript $(I,H(I),r)$ as is would be to impersonate the prover in an honest execution of the protocol. (This intuition is formalized in the proof of the theorem that follows.) By including a message m in the input to H, an accepting transcript $(I,H(I,m),r)$ thus constitutes a signature on m. Since $H(I,m)$ is redundant (as it can be computed from I and m), the actual signature is just (I,r).

The Fiat-Shamir transform is formalized as Construction 8.1.

Construction 8.1: The Fiat-Shamir transform

Let $\Pi = (\mathsf{Gen}, \mathscr{P}, \mathscr{V})$ be a canonical identification scheme where the verifier's challenges are chosen uniformly from Ω. Let $H : \{0,1\}^* \to \Omega$ be a hash function.

Key generation: Run $\mathsf{Gen}(1^k)$ to generate keys (pk, sk). These are the public and secret keys, respectively.

Signature generation: To sign message m using secret key sk, do:

- Run the prover algorithm $\mathscr{P}(sk)$ to generate an initial message I.
- Compute $c := H(I, m)$.
- Compute the appropriate response r to the "challenge" c using $\mathscr{P}(sk)$.

The signature is (I, r).

Signature verification: To verify the signature (I, r) on message m with respect to public key pk, proceed as follows:

- Compute $c := H(I, m)$.
- Accept the signature iff (pk, I, c, r) is an accepting transcript.

Theorem 8.1. *Let $\Pi = (\mathsf{Gen}, \mathscr{P}, \mathscr{V})$ be a canonical identification scheme that is secure against a passive attack. Then if H is modeled as a random oracle, the signature scheme Π' resulting from the Fiat-Shamir transform applied to Π is existentially unforgeable under an adaptive chosen-message attack.*

Proof. The intuition behind the proof is relatively straightforward, although there are some technical details to take care of. We will use an adversary A' attacking the signature scheme to construct an adversary A attacking the identification scheme. Adversary A is given a public key pk as well as access to the oracle $\mathsf{Trans}_{sk,pk}$, and interacts with an honest verifier $\mathscr{V}$ (recall by the discussion following Definition 8.2 that we can allow A to access the Trans oracle even during its interaction with $\mathscr{V}$). To get a feel for the main idea, assume for simplicity that A' acts in the following

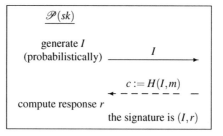

Fig. 8.3 "Collapsing" an identification scheme using the Fiat-Shamir transform.

way: it first requests signatures on (distinct) messages $m_1, \ldots, m_\ell$; next, it makes a single hash query $c := H(I, m')$ (where $m' \notin \{m_1, \ldots, m_\ell\}$); finally, it outputs the signature forgery (I, r) on the message m'.

To construct A, we must show two things: how it can simulate the signing queries made by A', and how it can use the forgery produced by A' to attack the identification scheme Π. For the "toy" example just mentioned these are both quite simple: In response to a request by A' for a signature on the message m_i, we have A proceed as follows:

1. Query $\mathsf{Trans}_{sk,pk}$ to obtain a transcript (I_i, c_i, r_i) of an execution of Π.
2. Return the signature (I_i, r_i) to A'. (By doing so, A implicitly sets $H(I_i, m_i) = c_i$.)

For each message m_i, the signature (I_i, r_i) returned by A is distributed *identically* to a signature that would be generated using signature scheme Π'. To see this, compare how the values (I_i, c_i, r_i) are generated in each case:

An execution of Π	An execution of Π'
1. Choose random coins ω_i.	1. Choose random coins ω_i.
2. $\mathcal{P}(sk; \omega_i)$ generates I_i.	2. $\mathcal{P}(sk; \omega_i)$ generates I_i.
3. $\mathcal{V}$ chooses random $c_i \leftarrow \Omega$.	3. $c_i := H(I_i, m_i)$.
4. $\mathcal{P}(sk, c_i; \omega_i)$ outputs r_i.	4. $\mathcal{P}(sk, c_i; \omega_i)$ outputs r_i.

The only difference is in step 3; however, since H is a random oracle, the c_i's are distributed identically in the two experiments. (Though there is no randomness explicit in step 3 of the execution of Π', the randomness is implicit in the initial, random selection of H from the set of all functions.) We conclude that, for this particular A', our adversary A is able to perfectly simulate all signing queries of A'.

We next describe how A can use the forgery of A' to successfully impersonate $\mathcal{P}$. When A' makes its hash query $H(I, m')$, then A simply sends the initial message I to the verifier $\mathcal{V}$. When $\mathcal{V}$ responds with a challenge c, then A responds to the hash query of A' with exactly this value. Finally, when A' outputs the signature forgery (I, r) on message m', the response r is forwarded by A to $\mathcal{V}$; see Figure 8.4. Note that

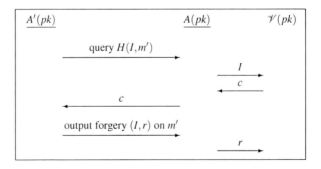

Fig. 8.4 A uses A' to impersonate the honest prover $\mathcal{P}$.

(1) c is uniformly distributed in Ω, exactly as the output of a random oracle should be; and (2) if $\mathsf{Vrfy}_{pk}(m',(I,r)) = 1$ then $(I,H(I,m'),r)$ is an accepting transcript and so $\langle A, \mathcal{V}(pk) \rangle = 1$; that is, A succeeds in impersonating $\mathcal{P}$ exactly when A' outputs a valid forgery.

We now generalize these ideas to construct an adversary A from *any* forger A'. Among the difficulties we must now deal with are that A does not know which hash query of A' will be used by A' to construct its forgery; also, A' may interleave its hash and signing queries in an arbitrary fashion.

So, let A' be a PPT adversary attacking Π'. We make a few simplifying assumptions without any loss of generality. First, we assume that A' makes any given hash query only once. For convenience, when a signature (I,r) on a message m is given to A' we also include the value $H(I,m)$; we may therefore assume that A' never queries $H(I,m)$ after receiving such a signature. We also require that if A' outputs the forgery (I,r) on a message m, then A' had previously asked the hash query $H(I,m)$. We will call this unique such query the *special* hash query. Let q_H be a polynomial upper bound on the number of hash queries made by A'.

We now describe out PPT adversary A attacking Π. Adversary A is given as input a public key pk, has access to the oracle $\mathsf{Trans}_{sk,pk}$, and interacts with a verifier $\mathcal{V}$. The first thing A does is guess at random an index $i \leftarrow \{1,\ldots,q_H\}$; this represents a guess as to the index of the special hash query (if any) that will be made by A'. Adversary A then runs $A'(pk)$, answering the oracle queries of A' as follows

Hash query $H(I,m)$: There are two cases:

- If this is the ith query to H, then A is (implicitly) guessing that this query will be the special hash query. A sends I to the honest verifier $\mathcal{V}$ with which it is interacting, and receives in return a challenge c. Then A returns c as the answer to this hash query.
  We will refer to this hash query as the *guessed query*.
- If this is not the ith hash query, then A simply chooses a random value $c \leftarrow \Omega$ and returns c as the answer to this query.

In either case, the value returned to A' is uniformly distributed in Ω as required.

Signing query $\mathsf{Sign}_{sk}(m)$: A queries its oracle $\mathsf{Trans}_{sk,pk}$ and obtains in return a transcript (I,c,r). If the hash value $H(I,m)$ was previously defined, then A aborts. If not, then A returns the signature (I,r) to A' (along with the hash value $H(I,m) = c$).

If A' outputs a valid forgery $(\hat{I},\hat{r})$ on a message $\hat{m}$, then A checks whether the guessed query $H(I,m)$ is equal to the special query $H(\hat{I},\hat{m})$. If not, then A aborts. Otherwise, A sends $\hat{r}$ to its verifier $\mathcal{V}$. Observe that as long as A does not abort during its execution, and its guess for the special query is correct, A succeeds in impersonating the honest prover. This is so because

- When the guessed query is equal to the special query, this means that A has sent $\hat{I}$ to its verifier, and received in return the challenge $c = H(\hat{I},\hat{m})$.
- $(\hat{I},\hat{r})$ is a valid signature on $\hat{m}$ exactly if $(\hat{I},c,\hat{r})$ is an accepting transcript.

A aborts if, in the process of answering a signing query for the message m, it happens that A obtains a transcript (I,c,r) for which the hash query $H(I,m)$ was already made by A'. Since the identification protocol Π is assumed to be canonical (and so the first message of the protocol is non-degenerate), this event occurs with only negligible probability.

Assuming A did not abort during its execution, it provides a perfect simulation for A'. Moreover, A's guess for the special query is correct with probability $1/q_H$ (and this event is independent of the event that A' outputs a valid forgery). We conclude that if A' succeeds in outputting a valid forgery with probability ε', then A succeeds in its impersonation attempt with probability exactly $\varepsilon = (\varepsilon' - \mu(k))/q(k)$ for some negligible function μ. Since identification scheme Π is secure by assumption, this completes the proof.

It is easy to see that, in general, the signature scheme Π' need not be *strongly* unforgeable. For example, say Π is a canonical identification scheme having the property that $(pk,I,c,r0)$ is an accepting transcript iff $(pk,I,c,r1)$ is; i.e., the last bit of $\mathscr{P}$'s response is ignored by $\mathscr{V}$. Then in the derived scheme Π', a valid signature (I,rb) on a message m can be changed to the different valid signature $(I,r\bar{b})$ on the same message. On the other hand, it can be shown (via a modification of the above proof) that this sort of attack is the only possibility. Thus, if Π has the property that for any pk,I,c there is (at most) a single r so that (pk,I,c,r) is an accepting transcript, then the derived scheme Π' is strongly unforgeable under an adaptive chosen-message attack.

A second observation, which proves quite useful in practice, is that for some specific identification schemes a more efficient variant of the Fiat-Shamir transform is possible. Specifically, say a canonical identification scheme Π has the property that given any public key pk, any challenge c, and any response r, it is possible to deterministically compute (in polynomial time) an initial message I such that (pk,I,c,r) is an accepting transcript. Then we can modify the signature scheme Π' so that the signature is (c,r) rather than (I,r). (Verification is done in the natural way; see Construction 8.2.) Since signatures computed according to this variant approach can be converted to signatures computed using the approach of Construction 8.1, and vice versa, existential unforgeability of this variant follows. (Note, however, that for strong unforgeability of this variant an additional assumption on the identification protocol is needed beyond the assumption required for strong unforgeability for Construction 8.1.)

8.2.2 Two Useful Criteria

In this section we will explore two useful criteria for identification schemes that, if satisfied, imply passive security. (We continue to assume canonical identification schemes, although most of what we say can be appropriately generalized if this is not the case.) The criteria given here are *sufficient* for passive security, but not *necessary*. Nevertheless, introducing these criteria has a three-fold benefit: (1) many

Construction 8.2: The Fiat-Shamir transform (variant)

Key generation: As before.

Signature generation: To sign message m using secret key sk, do:

- Run the prover algorithm $\mathscr{P}(sk)$ to generate an initial message I.
- Compute $c := H(I, m)$.
- Compute the appropriate response r to the "challenge" c using $\mathscr{P}(sk)$.

The signature is (c, r).

Signature verification: To verify the signature (c, r) on message m with respect to public key pk, do:

- Deterministically compute I such that (pk, I, c, r) is an accepting transcript. If no such A exists, reject.
- Accept the signature $c \overset{?}{=} H(I, m)$.
- Accept the signature iff (pk, I, c, r) is an accepting transcript.

---

identification schemes satisfy these criteria, and as such it is often convenient to be aware of this fact when considering other applications of those schemes; (2) introducing these criteria helps simplify and unify proofs of security for the identification schemes we will see later in the chapter (as well as many others that have been proposed in the literature); (3) as noted earlier in this chapter, the criteria defined here are extremely useful in other areas of cryptography (outside the realm of identification schemes), and so there is an independent benefit to studying them.

Informally, the two criteria are as follows:

Honest-verifier zero knowledge (HVZK)   There is an efficient probabilistic algorithm Sim that takes a public key pk and outputs transcripts that are indistinguishable from transcripts of honest executions of the identification protocol. While computational indistinguishability suffices, we remark that all the identification schemes considered in this chapter satisfy *perfect* HVZK whereby the transcripts output by Sim are identically distributed to the transcripts of real executions $\langle \mathscr{P}(SK), \mathscr{V}(pk) \rangle$.

Special soundness   Given public key pk (but not the associated secret key sk), it is computationally infeasible to output two accepting transcripts (I, c_1, r_1), (I, c_2, r_2) with the same first message but $c_1 \neq c_2$. Roughly speaking (we will see a more precise formulation of this below), this implies that for any PPT adversary trying to impersonate $\mathscr{P}$ the following holds: for any initial message I of the adversary's choosing, the adversary will only be able to respond correctly to at most one challenge (and so the adversary's probability of successful impersonation is at most $|\Omega|^{-1}$).

We now make these ideas more formal.

Definition 8.3. An identification protocol Π is **honest-verifier zero knowledge** (**HVZK**) if there exists a PPT algorithm Sim such that the following distributions are computationally indistinguishable:

$$\left\{ (pk,sk) \leftarrow \mathsf{Gen}(1^k) : \Big(sk,\, pk,\, \mathsf{Sim}(pk) \Big) \right\}$$

and

$$\left\{ (pk,sk) \leftarrow \mathsf{Gen}(1^k) : \Big(sk,\, pk,\, \mathsf{Trans}_{sk,pk} \Big) \right\}.$$

If the above distributions are *identical*, we say Π is **perfect** honest-verifier zero knowledge.

By a standard hybrid argument,[1] if Π is honest-verifier zero knowledge (resp., perfect honest-verifier zero knowledge) then the following is negligible (resp., zero) for any PPT adversary A:

$$\Big| \Pr\Big[(pk,sk) \leftarrow \mathsf{Gen}(1^k) : A^{\mathsf{Trans}_{sk,pk}(\cdot)}(pk) = 1 \Big]$$
$$- \Pr\Big[(pk,sk) \leftarrow \mathsf{Gen}(1^k) : A^{\mathsf{Sim}(pk)}(pk) = 1 \Big] \Big|$$

(where Sim is the algorithm guaranteed by Definition 8.3). Of course, one could simply define honest-verifier zero knowledge via the above equation (which is, in general, weaker than Definition 8.3); however, it will be much easier to prove honest-verifier zero knowledge using Definition 8.3.

Definition 8.4. An identification protocol Π satisfies **special soundness** if the following is negligible for all PPT algorithms A:

$$\Pr\left[\begin{array}{c|c} (pk,sk) \leftarrow \mathsf{Gen}(1^k) & c_1 \neq c_2 \\ (I,c_1,r_1,c_2,r_2) \leftarrow A(pk) & \text{and} \\ & (pk,I,c_1,r_1),\, (pk,I,c_2,r_2) \\ & \text{are both accepting transcripts} \end{array} \right].$$

The following result is quite natural, although the proof is a bit technical.

Theorem 8.2. *Assume canonical identification protocol Π is honest-verifier zero knowledge and satisfies special soundness. Then for any PPT adversary $A = (A_1, A_2)$ we have:*

$$\Pr\left[\begin{array}{c|c} (pk,sk) \leftarrow \mathsf{Gen}(1^k) & : \langle A_2(s), \mathscr{V}(pk) \rangle = 1 \\ s \leftarrow A_1^{\mathsf{Trans}_{sk,pk}(\cdot)}(pk) & \end{array} \right] - |\Omega_k|^{-1} \leq \mu(k)$$

for some negligible function $\mu(\cdot)$. In particular, if $|\Omega_k| = \omega(\mathrm{poly}(k))$ then Π is secure against a passive attack.

---

[1] The proof relies on the fact that, in Definition 8.3, indistinguishability holds *even given the real secret key sk*. This is one reason we formulated the definition as we did, even though in the literature the definition is often formulated as requiring indistinguishability of the transcripts only.

The theorem can be modified appropriately for the case when the space of challenges Ω depends on the public key.

Proof. Given a PPT adversary A attacking Π in an impersonation attack, we construct an algorithm B that violates special soundness. (B, as described, runs in expected polynomial time; using standard techniques, it can be modified to run in strict polynomial time.) The intuition is simple: B, given a public key pk, runs $A_1(pk)$. The Trans queries of A_1 are simulated by B using the procedure Sim for simulating executions of protocol Π (such a Sim exists since Π is honest-verifier zero knowledge). When A_1 is done, it passes some state s to A_2 that, informally, enables A_2 to impersonate $\mathscr{P}$ with "high" probability. Now, B runs A_2 with B playing the role of the verifier. The intuition here is that since A_2 succeeds with "high" probability (over random choice of challenge $c \in \Omega_k$), then with high probability B will be able to find two distinct challenges c_1, c_2 for which A_2 answers correctly on the same first message I (note A_2's initial message I is fixed before the challenge is sent by B). This exactly violates special soundness. See Figure 8.5 for a high-level depiction of the main idea.

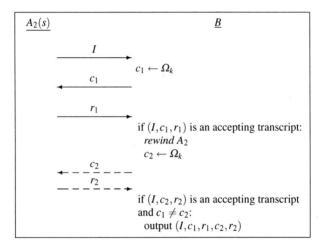

Fig. 8.5 Intuition as to how B violates special soundness using A_2.

For technical reasons, however, the proof is a bit more complicated. We formally define B in two stages. The first stage of B exactly follows the intuition outlined above, resulting in A_1 outputting state s that is then passed to A_2:

Algorithm B:
The algorithm is given a public key pk.

- Run $A_1(pk)$.
- To respond to a query $\mathsf{Trans}_{sk,pk}(\cdot)$ made by A_1, run $\mathsf{Sim}(pk)$ and give the result to A_1.

- When A_1 is done, it outputs state s. Run procedure Extract (below) using $A_2(s)$.
- If Extract outputs "success", then output (I, c_1, r_1, c_2, r_2) with $c_1 \neq c_2$ and (I, c_1, r_1), (I, c_2, r_2) both accepting transcripts.

The second stage of B, which we have called Extract above, uses $A_2(s)$ in an attempt to compute two accepting transcripts (I, c_1, r_1) and (I, c_2, r_2) with $c_1 \neq c_2$. In our description of the Extract algorithm, we implicitly assume that A_2 is deterministic; this is without loss of generality, as any random coins for A_2 could just as well be included in the state s. Thus, we may view $A_2(s)$ as follows: first, $A_2(s)$ outputs some initial message I; next, for any challenge $c \in \Omega_k$, running $A_2(s, c)$ results in some response r.

Algorithm Extract:
This algorithm interacts with $A_2(s)$.

- Run $A_2(s)$, which outputs some initial message I.
- Choose $c_1 \leftarrow \Omega_k$, and compute $r_1 := A_2(s; c_1)$.
- If (I, c_1, r_1) is not an accepting transcript, then output "fail" and abort
- Otherwise, for $c' \in \Omega_k \setminus \{c_1\}$ do:
  - Choose $c_2 \leftarrow \Omega_k \setminus \{c_1\}$ and compute $r_2 := A_2(s; c_2)$.
  - Also compute $r' := A_2(s; c')$.
  - If either (I, c_2, r_2) or (I, c', r') are accepting, output "success" and stop.
- Output "fail" and abort.

We analyze the behavior of B in the following claims.

Claim. B runs in expected polynomial time.

Proof. It is clear that the main routine of B runs in strict polynomial time (assuming A_1 does). What remains is to analyze the running time of Extract. Fix the algorithm $A_2(s)$ that Extract interacts with; we show that, regardless of the behavior of $A_2(s)$ the number of times Extract runs A_2 is polynomial in expectation. (Thus, since A_2 runs in strict polynomial time the overall running time of Extract is polynomial in expectation.) It suffices to show that the expected number of iterations of the inner *for* loop is polynomial. Fix any s, and let p be the number of challenges $c \in \Omega_k$ for which $A_2(s)$ responds correctly (i.e., for which $(A_2(s), c, A_2(s, c))$ is an accepting transcript). Then the probability that $A_2(s)$ succeeds in impersonating the prover is exactly $p/|\Omega_k|$, and so Extract enters the *for* loop with exactly this probability. Assuming Extract enters the *for* loop, there are two possibilities:

Case 1: $p = 1$. In this case, Extract will never output "success" but will instead run the maximum $|\Omega_k| - 1$ iterations of the *for* loop. But since it enters the *for* loop with probability only $1/|\Omega_k|$, the *expected* number of iterations of the *for* loop is only:

$$\frac{1}{|\Omega_k|} \cdot (|\Omega_k| - 1) < 1.$$

Case 2: $p > 1$. Then there are $p - 1 > 0$ challenges $c_2 \in \Omega_k \setminus \{c_1\}$ for which $A_2(s)$ responds correctly. The probability of selecting such a challenge, in any given iteration of the *for* loop, is $(p-1)/(|\Omega_k| - 1) > (p-1)/|\Omega_k|$. It follows that the expected number of iterations of the *for* loop is at most $|\Omega_k|/(p-1)$. Since Extract only enters the *for* loop with probability $p/|\Omega_k|$, the expected number of iterations of the *for* loop, overall, is at most:

$$\frac{p}{|\Omega_k|} \cdot \frac{|\Omega_k|}{p-1} \leq 2.$$

This completes the proof of the claim.

The next claim shows that if $A_2(s)$ succeeds with high probability (over random choice of challenge $c \in \Omega_k$), then Extract outputs "success" with similarly high probability.

Claim. Fix A_2 and s and let

$$\Pr[\langle A_2(s), \mathcal{V} \rangle = 1] = \rho$$

(the above probability is over $\mathcal{V}$'s choice of challenge only). Then $\mathsf{Extract}^{A_2(s)}$ outputs "success" with probability at least $\rho - |\Omega_k|^{-1}$.

Proof. When $\rho \leq |\Omega|^{-1}$ the claim is trivially true. For the case of $\rho > |\Omega_k|^{-1}$ we show something stronger: that Extract outputs "success" with probability *exactly* ρ. To see this, note that Extract enters the *for* loop with probability exactly ρ. But when $\rho > |\Omega_k|^{-1}$, there are at least two challenges (and hence at least one challenge $c_2 \in \Omega_k \setminus \{c_1\}$) for which $A_2(s)$ responds correctly. Since Extract does not exit the *for* loop until all challenges in $\Omega_k \setminus \{c_1\}$ have been tried (to see whether $A_2(s)$ responds correctly), it is clear that Extract outputs "success" in this case whenever it enters the *for* loop. It follows that if $\rho > |\Omega_k|^{-1}$, then Extract outputs "success" with probability exactly ρ.

To complete the proof, we show that if

$$\mathsf{Adv}(k) \stackrel{\text{def}}{=} \Pr\left[\begin{array}{c} (pk, sk) \leftarrow \mathsf{Gen}(1^k) \\ s \leftarrow A_1^{\mathsf{Trans}_{sk,pk}(\cdot)}(pk) \end{array} : \langle A_2(s), \mathcal{V}(pk) \rangle = 1 \right] - |\Omega_k|^{-1}$$

is not negligible, then the probability that B outputs "success" (and hence violates special soundness) is non-negligible as well. Special soundness of Π then implies that Adv must be negligible, as desired.[2]

---

[2] Technically speaking, we are not quite done since B runs in *expected* polynomial time rather than *strict* polynomial time. But the same ideas as in footnote 6 of Chapter 4 apply here as well.

So, assume Adv is not negligible, and define

$$\mathsf{Adv}'(k) \overset{\text{def}}{=} \Pr\left[\begin{array}{l}(pk, sk) \leftarrow \mathsf{Gen}(1^k)\\ s \leftarrow A_1^{\mathsf{Sim}(pk)}(pk)\end{array} : \langle A_2(s), \mathscr{V}(pk)\rangle = 1\right] - |\Omega_k|^{-1}$$

(the difference is that we have replaced A_1's access to Trans with access to Sim). Since Π is honest-verifier zero knowledge, it must be that $\left|\mathsf{Adv}(k) - \mathsf{Adv}'(k)\right|$ is negligible (see also the remark following Definition 8.3) and hence Adv' is *not* negligible. Next, note that we may write

$$\mathsf{Adv}'(k) = \mathbf{Exp}\left[\begin{array}{l}(pk, sk) \leftarrow \mathsf{Gen}(1^k)\\ s \leftarrow A_1^{\mathsf{Sim}(pk)}(pk)\end{array} : \Pr\left[\langle A_2(s), \mathscr{V}\rangle = 1\right]\right] - |\Omega_k|^{-1},$$

where the outer expectation is taken over an experiment that is identical to the first phase of B, and the inner probability is taken over random choice of the verifier's challenge. Using Claim 8.2.2 we thus have:

$$\Pr[B \text{ outputs "success"}]$$

$$= \mathbf{Exp}\left[\begin{array}{l}(pk, sk) \leftarrow \mathsf{Gen}(1^k)\\ s \leftarrow A_1^{\mathsf{Sim}(pk)}(pk)\end{array} : \Pr\left[\text{Extract outputs "success"}\right]\right]$$

$$\geq \mathbf{Exp}\left[\begin{array}{l}(pk, sk) \leftarrow \mathsf{Gen}(1^k)\\ s \leftarrow A_1^{\mathsf{Sim}(pk)}(pk)\end{array} : \Pr\left[\langle A_2(s), \mathscr{V}\rangle = 1\right] - |\Omega_k|^{-1}\right]$$

$$= \mathbf{Exp}\left[\begin{array}{l}(pk, sk) \leftarrow \mathsf{Gen}(1^k)\\ s \leftarrow A_1^{\mathsf{Sim}(pk)}(pk)\end{array} : \Pr\left[\langle A_2(s), \mathscr{V}\rangle = 1\right]\right] - |\Omega_k|^{-1}$$

$$= \mathsf{Adv}'(k).$$

We thus see that if Adv' is not negligible, then B violates special soundness with non-negligible probability. This completes the proof of the theorem.

An identification protocol satisfying the criteria of Theorem 8.2 is said to achieve *soundness* $1/|\Omega_k|$.

8.2.3 One-Time Signature Schemes without Random Oracles

In the following section we will demonstrate several identification protocols that are honest-verifier zero knowledge and also satisfy special soundness. By Theorem 8.2, this implies passive security for each of those protocols and so the Fiat-Shamir transform can be applied to each of them to obtain a construction of a CMA-secure signature scheme in the random oracle model. Here, we identify here a slight strengthening of the HVZK condition and then show how this (along with special soundness) gives a simple construction of a one-time *KMA*-secure signature scheme *without relying on random oracles*.

We first describe the construction. The public key of the signature scheme is simply a public key for the underlying identification scheme, along with an initial first message I. The message $m \in \Omega$ to be signed is interpreted as a "challenge"; to sign this message, the signer simply computes the appropriate response r. (Verification is done in the natural way.) See Construction 8.3 for a formal specification.

Construction 8.3: One-time signatures from canonical identification schemes

---

Let $\Pi = (\mathsf{Gen}, \mathscr{P}, \mathscr{V})$ be a canonical identification scheme with challenge space $\Omega = \{\Omega_k\}$. (If Ω depends on the public key, the following scheme can be modified appropriately.)

Key generation: Run $\mathsf{Gen}(1^k)$ to generate keys (pk, sk). Then run the prover algorithm $\mathscr{P}(sk)$ to generate an initial message I. The public key is (pk, I) and the secret key includes sk as well as the random coins used to generate I.

Signature generation: To sign message $m \in \Omega$, interpret $m \in \Omega$ as a challenge and compute the appropriate response r using the prover algorithm, secret key sk, and random coins used to generate I (which are part of the signer's secret key). The signature is r.

Signature verification: To verify signature r on message m with respect to public key (pk, I), simply check whether (pk, I, m, r) is an accepting transcript.

To analyze the security of this construction, we introduce a slightly stronger variant of HVZK called *special HVZK*. Recall that HVZK requires the existence of a simulator Sim that, on input a public key pk, outputs transcripts indistinguishable from transcripts of real executions of the identification protocol between $\mathscr{P}$ and $\mathscr{V}$. Special HVZK, roughly speaking, requires this simulator to work in a specific way: namely, Sim additionally takes as input a challenge $c \in \Omega$ and outputs a transcript *in which c is the challenge*. We define this notion formally only for the case in which the resulting transcript is *perfectly* indistinguishable from real transcripts.

Definition 8.5. Identification protocol Π is **special honest-verifier zero knowledge** if there exists a PPT algorithm Sim such that the following distributions are identical:

$$\left\{ \begin{array}{l} (pk, sk) \leftarrow \mathsf{Gen}(1^k); c \leftarrow \Omega; \\ (I, r) \leftarrow \mathsf{Sim}(pk, c) \end{array} : \left(sk, pk, I, c, r \right) \right\}$$

and

$$\left\{ \begin{array}{l} (pk, sk) \leftarrow \mathsf{Gen}(1^k); c \leftarrow \Omega; \\ (I, s) \leftarrow \mathscr{P}(sk); \ r \leftarrow \mathscr{P}(s, c) \end{array} : (sk, pk, I, c, r) \right\}.$$

(In the above, s represents the prover's state.)

Note that if Π is special HVZK then it is also HVZK; given a special-HVZK simulator Sim we can construct an HVZK simulator Sim$'$ by simply choosing random $c \in \Omega$ and then running Sim(pk, c). We remark also that, for our application here, it would suffice in Definition 8.5 for indistinguishability to hold without being given sk (but in that case the definition would be incomparable to HVZK).

Theorem 8.3. *If Π is special honest-verifier zero knowledge and satisfies special soundness, then Construction 8.3 yields a one-time signature scheme that is existentially unforgeable under a known-message attack.*

Proof. The proof is straightforward, and so we only sketch it here. Given a PPT adversary A attacking the signature scheme, we construct a PPT adversary A' violating special soundness. Adversary A', given as input a public key pk, works as follows: first, it runs A to obtain message m (recall that in a known-message attack A must decide on its message before obtaining the public key). Then, it runs Sim(pk, c) to obtain (I, r). Finally, it returns to A the public key (pk, I) and the signature r on the message m. Note that both the public key and the signature are distributed exactly as in a real experiment involving A and the signer.

If A outputs a valid signature forgery r' on a message $m' \neq m$, then A' exactly violates special soundness of Π. The theorem follows.

Although the level of security obtained using Construction 8.3 is relatively weak, schemes designed using this methodology can be very efficient and thus highly useful, e.g., for application in Construction 1.2.

We conclude this section with two remarks on the proof:

- As in the case for signature schemes resulting from the Fiat-Shamir transform, signatures built using Construction 8.3 need not be *strongly* unforgeable. However, if Π has the property that for any pk, I, c there is (at most) a single r so that (pk, I, c, r) is an accepting transcript, then applying Construction 8.3 *does* yield a strongly unforgeable scheme.
- It suffices for the identification protocol to be HVZK (rather than special HVZK) in order to prove security of signatures resulting from Construction 8.3 under a *random* message attack.

8.3 Some Secure Identification Schemes

In order to apply the Fiat-Shamir transform and obtain concrete signature schemes, it remains only to show examples of canonical identification protocols that are both honest-verifier zero knowledge and satisfy special soundness (and also have exponentially large challenge spaces Ω). In fact, all the identification schemes we show will satisfy *special* HVZK and so Construction 8.3 can also be applied.

On parallel repetition. Before presenting any concrete examples of protocols, a brief remark is in order regarding the *parallel composition* of identification schemes.

To be precise, let $\Pi = (\text{Gen}, \mathscr{P}, \mathscr{V})$ be an identification protocol, let $t = t(k)$ be polynomial, and consider the protocol $\Pi^t = (\text{Gen}, \mathscr{P}^t, \mathscr{V}^t)$ in which $\mathscr{P}^t$ and $\mathscr{V}^t$ are obtained via t-fold (parallel) repetition of $\mathscr{P}$ and $\mathscr{V}$, respectively. (Note that Gen is unchanged, and so $\mathscr{P}^t$ and $\mathscr{V}^t$ execute parallel repetitions with respect to a single public key output by Gen.) It is not hard to see that if Π is (special) honest-verifier zero knowledge and satisfies special soundness, then Π^t does as well. If the challenge space for Π is Ω_k, then the challenge space for Π^t is $\Omega_k^{t(k)}$. We conclude (via application of Theorem 8.2) that if Π achieves soundness $\varepsilon(k)$ then Π^t achieves soundness $\varepsilon(k)^{t(k)}$. Parallel repetition therefore provides an *exponential* improvement to the soundness. A corollary is that, at least from a theoretical perspective, it suffices to design protocols achieving constant soundness (say, soundness $1/2$).

The above arguments extend to the case when the same protocol is repeated in parallel using different public keys, and also to the more general case when different protocols are executed in parallel.

In the sections that follow, we show a number of identification protocols satisfying the criteria required by Theorem 8.2. Many examples of such protocols have been proposed in the literature, and it is not our intention to provide an exhaustive survey here. Instead, we will focus on a few well-known, representative examples. We remark also that various efficiency improvements for the schemes we present are possible; our aim is to present the schemes in as straightforward and as simple a manner as possible.

8.3.1 The Fiat-Shamir Scheme

The first scheme we describe is also one of the first to have been proposed and proven secure. We start by presenting a "basic" version, due to Goldwasser, Micali, and Rackoff, and then discuss in detail a more efficient variation given by Fiat and Shamir. Both protocols rely for their security on the hardness of computing square roots modulo N (where N is a product of two primes); by Theorem 2.2, this is equivalent to basing security on the hardness of factoring.

The basic scheme is given as Construction 8.4. One can easily check that the verifier always accepts an honest execution of the protocol. We will prove next that, assuming the hardness of factoring, the protocol is honest-verifier zero knowledge and achieves special soundness; we conclude that the identification scheme has soundness $1/2$. Repeating the protocol sufficiently-many times in parallel (see the earlier discussion) allows us to achieve any desired level of security.

Theorem 8.4. *The Goldwasser-Micali-Rackoff identification scheme is (perfect) special honest-verifier zero knowledge. Furthermore, if factoring is hard relative to* GenModulus *then it also satisfies special soundness.*

Proof. We first prove that the Goldwasser-Micali-Rackoff (GMR) scheme is (perfect) special honest-verifier zero knowledge. Consider the following algorithm Sim:

Construction 8.4: The Goldwasser-Micali-Rackoff identification scheme

---

Let GenModulus be a PPT algorithm that, on input 1^k, outputs a modulus N along with two (distinct) k-bit primes p, q with $N = pq$.

Key generation: Run $(N, p, q) \leftarrow \mathsf{GenModulus}(1^k)$. Choose a random $x \in \mathbb{Z}_N^*$ and then set $y := x^2 \bmod N$. The public key is (N, y) and the secret key is x.

Prover's initial message: Choose $s \leftarrow \mathbb{Z}_N^*$ and send $I := s^2 \bmod N$.

Verifier's challenge: Choose and send challenge $c \leftarrow \{0, 1\}$.

Prover's response: Send $r := x^c \cdot s \bmod N$. (Here, x is taken from the secret key and s is the randomness used to generate the first-round message.)

Acceptance criterion: Accept transcript (I, c, r) with respect to the public key (N, y) if either (1) $c = 0$ and $r^2 = I \bmod N$, or (2) $c = 1$ and $r^2 = y \cdot I \bmod N$. (It is important also to verify that $I, r \in \mathbb{Z}_N^*$ and, in particular, $I \neq 0$.)

> **Algorithm** Sim:
> The algorithm is given a public key (N, y) in addition to a challenge $c \in \{0, 1\}$.
>
> - Choose $r \leftarrow \mathbb{Z}_N^*$.
> - Output the transcript $(r^2/y^c \bmod N, c, r)$.

To see that this perfectly simulates a real execution of the protocol for any given public key (N, y), any[3] secret key x, and any given challenge c we look — in each case — at the distribution of r (conditioned on the given values), and then at the distribution of I conditioned on r (and the given values). In a real execution of the protocol, r is uniformly distributed in $\mathbb{Z}_N^*$ regardless of the choice of c. (When $c = 0$ then $r = s$ and so this is clear; when $c = 1$ then $r = sx \bmod N$, but since s is uniform sx is uniform for any x.) Then I is the unique value satisfying $I = r^2/y^c \bmod N$. In a simulated transcript, r and I are chosen from identical distributions by construction.

Special soundness follows easily from the following observation: given two accepting transcripts $(I, 0, r_0)$ and $(I, 1, r_1)$ with respect to some public key (N, y), it is possible to efficiently compute a square root of y modulo N. (This implies special soundness, since then any PPT algorithm violating special soundness with non-negligible probability can be used to compute square roots of random elements of $\mathbb{Z}_N^*$ with non-negligible probability for N output by GenModulus. By Theorem 2.2, this contradicts the assumed hardness of factoring relative to GenModulus.) To see this, note simply that by definition of the acceptance criterion we have

$$r_0^2 = I = r_1^2/y \bmod N$$

---

[3] Recall that each public key has four possible associated secret keys.

with $r_0, r_1 \in \mathbb{Z}_N^*$, and so $r_1/r_0 \bmod N$ is a square root of y modulo N.

As we have already noted, by running the GMR protocol in parallel k times we can amplify the soundness from $1/2$ to 2^{-k}. Better communication complexity (at the expense of a longer public key) can be achieved using an approach due to Fiat and Shamir; see Construction 8.5 (and also Figure 8.6).

Construction 8.5: The Fiat-Shamir identification scheme

Let GenModulus be a PPT algorithm that, on input 1^k, outputs a modulus N along with two (distinct) k-bit primes p, q with $N = pq$.

Key generation: Compute $(N, p, q) \leftarrow \mathsf{GenModulus}(1^k)$. Then choose $x_1, \ldots, x_k \leftarrow \mathbb{Z}_N^*$ and set $y_i := x_i^2 \bmod N$ for all i. The public key is $(N, y_1, \ldots, y_k)$ and the secret key is $(x_1, \ldots, x_k)$.

Prover's initial message: Choose $s \leftarrow \mathbb{Z}_N^*$ and send $I := s^2 \bmod N$.

Verifier's challenge: Choose and send challenge $c \leftarrow \{0,1\}^k$.

Prover's response: Parse the challenge c as a sequence of bits $c_1, \ldots, c_k$. Send

$$r := s \cdot \prod_{i=1}^{k} x_i^{c_i} \bmod N = s \cdot \prod_{i|c_i=1} x_i \bmod N,$$

where the x_i are from the secret key and s is the randomness used in the first round.

Acceptance criterion: Accept transcript (I, c, r) (where $c = c_1 \cdots c_k$) with respect to the public key $(N, y_1, \ldots, y_k)$ iff

$$r^2 \overset{?}{=} I \cdot \prod_{i|c_i=1} y_i \bmod N.$$

(It is important also to verify that $I, r \in \mathbb{Z}_N^*$ and, in particular, $I \neq 0$.)

It is not difficult to verify that the honest prover is always accepted since, in an honest execution:

$$r^2 = \left(s \cdot \prod_{i|c_i=1} x_i \right)^2 = s^2 \cdot \prod_{i|c_i=1} x_i^2 = I \cdot \prod_{i|c_i=1} y_i \bmod N.$$

Theorem 8.5. *The Fiat-Shamir identification scheme is (perfect) special honest-verifier zero knowledge. Furthermore, if factoring is hard relative to* GenModulus *then it also satisfies special soundness.*

Proof. We first prove that the scheme is perfect special honest-verifier zero knowledge. Consider the following algorithm Sim:

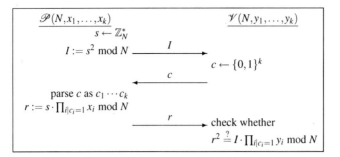

Fig. 8.6 An execution of the Fiat-Shamir identification scheme.

Algorithm Sim:
The algorithm is given a public key $(N, y_1, \ldots, y_k)$ and a challenge $c = c_1 \cdots c_k \in \{0, 1\}^k$.

- Choose $r \leftarrow \mathbb{Z}_N^*$.
- Set $I := r^2 \cdot \left(\prod_{i|c_i=1} y_i \right)^{-1} \bmod N$.
- Output the transcript (I, c, r).

To see that this perfectly simulates a real execution of the protocol for any public key $(N, y_1, \ldots, y_k)$, any secret key $(x_1, \ldots, x_k)$, and any challenge c we look — in each case — at the distribution of r conditioned on the given values, and then at the distribution of I conditioned on the given values and r. In an honest execution of the protocol, $r = s \cdot \prod_{i|c_i=1} x_i \bmod N$ for a uniformly distributed value $s \in \mathbb{Z}_N^*$ and so r is uniform regardless of c and $(x_1, \ldots, x_k)$. Then I is the unique value satisfying $I = r^2 \cdot \left(\prod_{i|c_i=1} y_i \right)^{-1} \bmod N$. In simulated transcripts r and I are distributed in the same manner, by construction.

Special soundness is a bit more involved.[4] We will show how to use any PPT algorithm which violates special soundness to factor moduli N output by GenModulus. Assume a PPT algorithm A for which

$$\Pr \left[\begin{array}{l} N \leftarrow \mathsf{GenModulus}(1^k); \\ x_1, \ldots, x_k \leftarrow \mathbb{Z}_N^*; \\ \forall i : y_i := x_i^2 \bmod N; \\ pk = (N, y_1, \ldots, y_k); \\ (I, c_1, r_1, c_2, r_2) \leftarrow A(pk) \end{array} \middle\vert \begin{array}{c} c_1 \neq c_2 \\ \text{and} \\ (pk, I, c_1, r_1), (pk, I, c_2, r_2) \\ \text{are both accepting transcripts} \end{array} \right]$$

is not negligible. Consider the following PPT algorithm F which takes as input a modulus N output by GenModulus: algorithm F first chooses random values $x_1, \ldots, x_k \leftarrow \mathbb{Z}_N^*$, and then sets $y_i := x_i^2 \bmod N$ for all i. Next, F runs A on input

---

[4] In contrast to the GMR scheme, a violation of special soundness here does not directly enable computation of the prover's secret key; for this reason, the proof of special soundness is somewhat more involved.

$(N, y_1, \ldots, y_k)$. Note that the input provided to A by F is distributed identically to a real public key in the Fiat-Shamir scheme, and so A violates special soundness with the same probability. Assuming A outputs (I, c_1, r_1, c_2, r_2) with $c_1 \neq c_2$ and where (I, c_1, r_1), (A, c_2, r_2) are both accepting transcripts, F proceeds as follows. By definition of the acceptance criterion,

$$r_1^2 \cdot \prod_{i=1}^{k} y_i^{-c_{1,i}} = I = r_2^2 \cdot \prod_{i=1}^{k} y_i^{-c_{2,i}} \bmod N,$$

where $c_{1,i}$ (resp., $c_{2,i}$) refers to the ith bit of c_1 (resp., c_2). We also have $r_1, r_2 \in \mathbb{Z}_N^*$. Let $\Delta c_i \overset{\text{def}}{=} c_{1,i} - c_{2,i}$. It follows that

$$(r_1/r_2)^2 = \prod_{i=1}^{k} y_i^{\Delta c_i} = \left(\prod_{i=1}^{k} x_i^{\Delta c_i} \right)^2 \bmod N,$$

and so F knows two square roots $R \overset{\text{def}}{=} (r_1/r_2)$ and $X \overset{\text{def}}{=} \prod_{i=1}^{k} x_i^{\Delta c_i}$ of the same value $Y \overset{\text{def}}{=} X^2 = R^2 \bmod N$. Note, however, that from the point of view of A the value X is a *random* square root of Y. (This is so since the x_i are chosen uniformly at random from $\mathbb{Z}_N^*$, the Δc_i lie in $\{-1, 0, 1\}$, and at least one Δc_i is non-zero.) As in Theorem 2.2, then, $R \neq \pm X \bmod N$ with probability $1/2$; furthermore, when this occurs F can compute the factorization of N. We conclude that if A indeed violates special soundness with non-negligible probability, then F factors moduli N output by GenModulus with non-negligible probability — but this contradicts the assumed hardness of factoring relative to GenModulus.

8.3.2 The Guillou-Quisquater Scheme

We now show an identification scheme introduced by Guillou and Quisquater whose security against passive attacks is based on the RSA assumption. See Construction 8.6 (and also Figure 8.7).

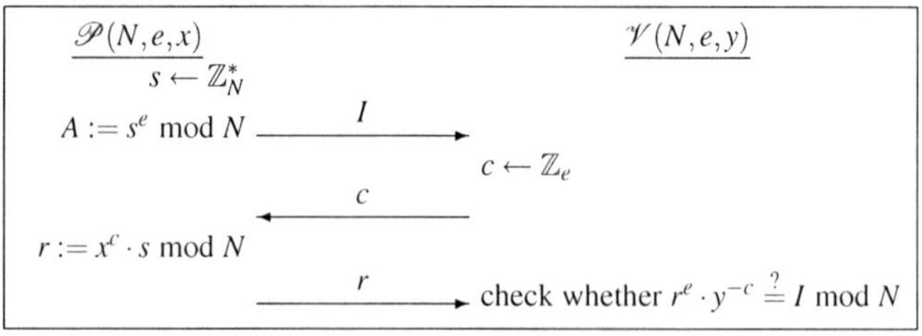

Fig. 8.7 An execution of the Guillou-Quisquater identification scheme.

Construction 8.6: The Guillou-Quisquater (GQ) Identification Scheme

Let GenRSA be as in Chapter 4.

Key generation: Compute $(N,e,d) \leftarrow \mathsf{GenRSA}(1^k)$, where e is prime. Then choose $x \leftarrow \mathbb{Z}_N^*$ and set $y := x^e \bmod N$. The public key is (N,e,y) and the secret key is x.

Prover's initial message: Choose $s \leftarrow \mathbb{Z}_N^*$ and send $I := s^e \bmod N$.

Verifier's challenge: Choose and send challenge $c \leftarrow \mathbb{Z}_e$.

Prover's response: Send the response $r := x^c \cdot s \bmod N$, where x is taken from the secret key and s is the randomness used in the first round.

Acceptance criterion: Accept transcript (I,c,r) with respect to the public key (N,e,y) iff

$$r^e \cdot y^{-c} \stackrel{?}{=} I \bmod N.$$

(It is important also to verify that $I, r \in \mathbb{Z}_N^*$ and, in particular, $I \neq 0$.)

In an honest execution of the Guillou-Quisquater protocol, the verifier always accepts since

$$r^e \cdot y^{-c} = (x^c \cdot s)^e \cdot y^{-c} = y^c \cdot s^e \cdot y^{-c} = s^e = I \bmod N.$$

Theorem 8.6. *The Guillou-Quisquater identification scheme is (perfect) special honest-verifier zero knowledge. Furthermore, if the RSA problem is hard relative to* GenRSA *then is also satisfies special soundness.*

Proof. We first prove that the GQ scheme achieves perfect special HVZK. Consider the following algorithm Sim:

> **Algorithm** Sim:
> The algorithm is given a public key (N,e,y) in addition to a challenge $c \in \mathbb{Z}_e$.
>
> - Choose $r \leftarrow \mathbb{Z}_N^*$.
> - Set $I := r^e \cdot y^{-c} \bmod N$.
> - Output the transcript (I,c,r).

To see that this perfectly simulates a real execution of the protocol for any given public key (N,e,y) (note that this fixes the secret key) and challenge c, we look — in each case — at the distribution of r conditioned on the given values, and then at the distribution of I conditioned on the given values and r. In a real execution of the protocol, r is uniformly distributed in $\mathbb{Z}_N^*$ (regardless of c) since $r = x^c \cdot s \bmod N$ for a uniformly distributed value $s \in \mathbb{Z}_N^*$. Then I is the unique value satisfying $I = r^e \cdot y^{-c} \bmod N$. In simulated transcripts, both r and I are chosen from these same distributions by construction.

Special soundness follows easily from the following observation: given two accepting transcripts (I, c_1, r_1) and (I, c_2, r_2) (with $c_1 \neq c_2$ and $c_1, c_2 \in \mathbb{Z}_e$) with respect to an arbitrary public key (N, e, y), it is possible to efficiently compute $y^{1/e} \bmod N$. (This implies special soundness, as any PPT algorithm violating special soundness with non-negligible probability can thus be used to solve the RSA problem with respect to GenRSA.) To see that this is true, note that by definition of the acceptance criterion we have

$$r_1^e \cdot y^{-c_1} = I = r_2^e \cdot y^{-c_2} \bmod N$$

with $r_1, r_2 \in \mathbb{Z}_N^*$, and so $(r_1/r_2)^e = y^{c_1 - c_2} \bmod N$. Now, since $c_1, c_2 \in \mathbb{Z}_e$ the absolute value of their difference is less than e; since e is prime,

$$\gcd(e, |c_1 - c_2|) = 1$$

and we may apply Lemma 4.1 to compute $y^{1/e} \bmod N$.

The GQ protocol achieves soundness $1/e$. To achieve passive security, then, we need to choose e large enough so that $1/e$ is negligible. (Alternately, we may use a smaller e and parallel repetition.)

8.3.3 The Micali/Ong-Schnorr Scheme

The next scheme we consider is secure under the assumption that factoring *Blum integers* is hard (cf. Section 2.2.1). The scheme is more efficient than the Fiat-Shamir scheme (whose security is based on essentially the same assumption), but we have deferred its presentation to now because of its similarities to the Guillou-Quisquater scheme.

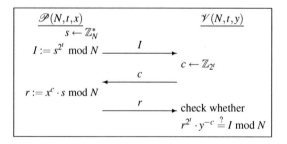

Fig. 8.8 An execution of the Micali/Ong-Schnorr identification scheme.

As usual, it is easy to check that correctness always holds:

$$r^{2^t} \cdot y^{-c} = (x^c \cdot s)^{2^t} \cdot y^{-c} = y^c \cdot s^{2^t} \cdot y^{-c} = I \bmod N.$$

Construction 8.7: The Micali/Ong-Schnorr Identification Scheme

Let GenModulus be a PPT algorithm that, on input 1^k, outputs a modulus N along with two (distinct) k-bit primes p, q with $N = pq$ and $p = q = 3 \bmod 4$. Fix some parameter $t = \omega(\log k)$.

Key generation: Compute $(N, p, q) \leftarrow$ GenModulus(1^k). Choose $x \leftarrow \mathbb{Z}_N^*$ and compute $y := x^{2^t} \bmod N$. The public key is (N, y) and the secret key is x.

Prover's initial message: Choose $s \leftarrow \mathbb{Z}_N^*$ and send $I := s^{2^t} \bmod N$.

Verifier's challenge: Choose and send challenge $c \leftarrow \mathbb{Z}_{2^t}$.

Prover's response: Send the response $r := x^c \cdot s \bmod N$, where x is taken from the secret key and s is the randomness used in the first round.

Acceptance criterion: Accept transcript (I, c, r) with respect to the public key (N, y) iff

$$r^{2^t} \cdot y^{-c} \stackrel{?}{=} I \bmod N.$$

(It is important also to verify that $I, r \in \mathbb{Z}_N^*$ and, in particular, $I \neq 0$.)

The similarity to the GQ scheme is clear. On the other hand, the *analysis* is more complex. The issue is this: from two accepting transcripts (I, c_1, r_1), (I, c_2, r_2) (with $c_1 \neq c_2$), one can compute (as in the proof of special soundness for the Guillou-Quisquater scheme):

$$(r_1/r_2)^{2^t} = y^{c_1 - c_2} \bmod N. \tag{8.2}$$

But now $\gcd(2^t, c_1 - c_2)$ is *not* necessarily equal to 1, and so we cannot necessarily compute $y^{1/2^t} \bmod N$. Instead, using the fact that N is a Blum integer, we show that Equation (8.2) enables computation of a *square root* of y. As in Theorem 2.2, this contradicts the assumed hardness of factoring N.[5] We now proceed with the details.

Theorem 8.7. *The Micali/Ong-Schnorr scheme is (perfect) special honest-verifier zero knowledge. Furthermore, if factoring is hard for* GenModulus *then it also satisfies special soundness.*

Proof. A proof that the scheme is honest-verifier zero knowledge follows along similar lines to the proof for the case of the Guillou-Quisquater scheme, and so we omit it. Before proving special soundness, we recall a key property of Blum integers (cf. Section 2.2.1: if N is a Blum integer, then squaring modulo N is a permutation on the set QR_N of quadratic residues modulo N. Thus, if $x, y \in QR_N$ and $x^2 = y^2 \bmod N$

---

[5] As in the case of the Fiat-Shamir scheme, then, a violation of special soundness does not enable computation of the prover's secret key (but can still be shown to be infeasible).

we can conclude that $x = y \bmod N$; more generally, if $x^{2^\ell} = y^{2^\ell} \bmod N$ then $x = y \bmod N$. We stress that this implication is valid only when $x, y \in \mathsf{QR}_N$.

With the above in mind, we first show that given two accepting transcripts (I, c_1, r_1) and (I, c_2, r_2) (with $c_1 \neq c_2$ and $c_1, c_2 \in \mathbb{Z}_{2^t}$) it is possible to compute a square root of y modulo N. As in the Guillou-Quisquater scheme, we may derive Equation (8.2); assume without loss of generality that $c_1 > c_2$ and let $\Delta = c_1 - c_2$. Compute integers ℓ, m (with $m \geq 1$ odd) such that $\Delta = 2^\ell \cdot m$; note that this can be done easily, and does not require the full factorization of Δ. Note further that since $c_1, c_2 \in \mathbb{Z}_{2^t}$ we have $\Delta < 2^t$ and so $\ell < t$.

Re-writing Equation (8.2), we have:

$$((r_1/r_2)^{2^{t-\ell}})^{2^\ell} = (y^m)^{2^\ell} \bmod N.$$

The key observation is that $y^m \in \mathsf{QR}_N$ (since $y \in \mathsf{QR}_N$) and furthermore it is also the case that $(r_1/r_2)^{2^{t-\ell}} \in \mathsf{QR}_N$ (since $t - \ell \geq 1$). By what we have said above, we conclude that

$$((r_1/r_2)^{2^{t-\ell-1}})^2 = y^m \bmod N$$

(using the fact that N is a Blum integer). Since $\gcd(2, m) = 1$, we may apply Lemma 4.1 to compute a square root of y.

8.3.4 The Schnorr Scheme

The final identification scheme we present in this section is due to Schnorr, and is based on the hardness of the discrete logarithm problem. For concreteness and simplicity, we describe the protocol based on some fixed group $\mathbb{G}$ of (publicly known) prime order q; in general, of course, the size of the group will have to grow with the security parameter k and so we need an infinite family of such groups along with an efficient algorithm generating (descriptions of) groups of the required form. A description of the scheme follows (see also Construction 8.8):

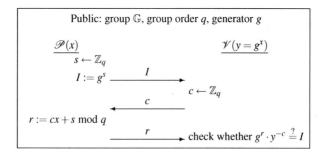

Fig. 8.9 An execution of the Schnorr identification scheme.

Construction 8.8: The Schnorr identification scheme

Assume a group $\mathbb{G}$ of prime order q and a generator $g \in \mathbb{G}$ are publicly known (otherwise, they can also be generated during key generation and included with the public key).

Key generation: Choose random $x \leftarrow \mathbb{Z}_q$ and set $y := g^x$. The public key is $y \in \mathbb{G}$ and the secret key is $x \in \mathbb{Z}_q$.

Prover's initial message: Choose $s \leftarrow \mathbb{Z}_q$ and send $I := g^s$.

Verifier's challenge: Choose and send challenge $c \leftarrow \mathbb{Z}_q$.

Prover's response: Send the response $r := cx + s \bmod q$, where x is the secret key and s is the randomness used in the first round.

Acceptance criterion: Accept transcript (A, c, r) with respect to the public key y iff

$$g^r \cdot y^{-c} \stackrel{?}{=} I.$$

(In general, one must also verify that $I \in \mathbb{G}$ and $c \in \mathbb{Z}_q$.)

Correctness is easily verified.

Theorem 8.8. *The Schnorr identification scheme is (perfect) special honest-verifier zero knowledge. Furthermore, if the discrete logarithm problem is hard in $\mathbb{G}$ then it also satisfies special soundness.*

Proof. We first prove that the Schnorr scheme achieves perfect special HVZK. Consider the following algorithm Sim:

> **Algorithm** Sim:
> The algorithm is given a public key y and a challenge $c \in \mathbb{Z}_q$.
>
> - Choose $r \leftarrow \mathbb{Z}_q$.
> - Set $I := g^r \cdot y^{-c}$.
> - Output the transcript (I, c, r).

To see that this perfectly simulates a real execution of the protocol for any given public key y (which fixed the secret key x) and challenge c, we look — in each case — at the distribution of r conditioned on the given values, and then at the distribution of I conditioned on the values of the given values and r. In a real execution of the protocol, r is uniformly distributed in $\mathbb{Z}_q$ (regardless of c) because $r = cx + s \bmod q$ for a uniformly distributed value $s \in \mathbb{Z}_q$. Then I is the unique value satisfying $I = g^r \cdot y^{-c}$. In simulated transcripts, r and I have the same distributions by construction.

Special soundness follows easily from the following observation: given two accepting transcripts (I, c_1, r_1) and (I, c_2, r_2) (with $c_1 \neq c_2$ and $c_1, c_2 \in \mathbb{Z}_q$) with respect to a public key y, it is possible to efficiently compute $\log_g y$. (To see that this implies special soundness, note that any PPT algorithm violating special soundness with non-negligible probability can thus be used to compute discrete logarithms of a random group element $y \in \mathbb{G}$ with non-negligible probability.) To see this, note that by definition of the acceptance criterion we have

$$g^{r_1} y^{-c_1} = I = g^{r_2} y^{-c_2}$$

and so $g^{r_1 - r_2} = y^{c_1 - c_2}$ and

$$g^{(r_1 - r_2)/(c_1 - c_2)} = y.$$

We conclude that $\log_g y = (r_1 - r_2) \cdot (c_1 - c_2)^{-1} \bmod q$, which can be computed efficiently (note that $c_1 - c_2 \neq 0 \bmod q$ and so the desired inverse modulo q exists).

8.4 Further Reading

Identification schemes have a long history, and we have only covered here what is necessary for our treatment of signature schemes. The reader interested in identification schemes *per se* is referred to the references cited here, as well as other works (e.g., [93]), for a full discussion of stronger definitions of security and how they can be achieved.

Canonical identification schemes as we have described them here are also known as *sigma protocols* (Σ-protocols). This abstract notion, as well as the definitions of (special) HVZK and special soundness, were first proposed in [32]; see there for further discussion and applications.

The Fiat-Shamir transform was first presented in [47]. Our proof here follows that of Abdalla et al. [1], though earlier proofs of slightly weaker results were also known [96, 92]. The proof relies heavily on the random oracle model, and it remains an active area of research to give a proof of the Fiat-Shamir transform (or a suitable variant) in the standard model, or to show the insecurity of practical schemes derived using the Fiat-Shamir transform. For some negative results indicating limitations on what can be proved regarding the Fiat-Shamir transform in general, see [43, 58].

The weaker transformation from identification protocols to KMA-secure one-time signature schemes (Construction 8.3) is implicit in work of Cramer and Damgård [33], where it was then used to construct an existentially unforgeable signature scheme. The idea has also been formalized as *chameleon signatures* [74], which have several applications. See the work of [103] for some particularly efficient instantiations.

Construction 8.4 is implicit in the work of Goldwasser, Micali, and Rackoff [60], though it was presented in a somewhat different context there. Fiat and Shamir [47] gave the improved Construction 8.5 and analyzed it as an identification scheme. A slight generalization of the Fiat-Shamir scheme was studied in later work by Feige, Fiat, and Shamir [46]. Construction 8.6 was presented by Guillou and Quisquater [63].

Micali introduced a signature scheme derived from an identification scheme much like Construction 8.7 in an unpublished manuscript [82]; the scheme was first published by Ong and Schnorr [94]. Several variants and extensions of this construction have been given in the literature [91, 69, 104, 49]. Construction 8.8 is due to Schnorr [101].

In contrast to Chapter 7, here we have not focused much on the concrete security bounds of our proofs of security. In fact, the reader may notice that the concrete security bound guaranteed by our proof of Theorem 8.1 is quite bad; it is open whether this can be improved for specific schemes. An alternative to the Fiat-Shamir transform was suggested by Micali and Reyzin [84]; their transform yields a tighter concrete security analysis, but only applies to restricted types of identification schemes. Work of Goh et al. [54] introduces signature schemes that can be viewed as being derived from some *specific* identification schemes; their schemes have tight reductions to the computational or decisional Diffie-Hellman assumptions.

A notable omission from this book is the Digital Signature Algorithm/Digital Signature Standard (DSA/DSS), a US government standard [90]. Despite its importance, we have made a conscious decision to leave a discussion of DSS out of this book because existing proofs of security for DSS (or variants thereof) require proof techniques beyond the scope of this book. The interested reader is referred to [21, 107, 77, 22] or the analyses of DSA/ECDSA posted on the CRYPTREC website.[6]

---

[6] http://www.cryptrec.go.jp/english/estimation.html

References

1. M. Abdalla, J. H. An, M. Bellare, and C. Namprempre. From identification to signatures via the Fiat-Shamir transform: Necessary and sufficient conditions for security and forward-security. *IEEE Transactions on Information Theory*, 54(8):3631–3646, 2008.
2. J.H. An, Y. Dodis, and T. Rabin. On the security of joint signature and encryption. In *Advances in Cryptology — Eurocrypt 2002*, volume 2332 of *LNCS*, pages 83–107. Springer, 2002.
3. B. Barak and M. Mahmoody-Ghidary. Lower bounds on signatures from symmetric primitives. In *48th Annual Symposium on Foundations of Computer Science (FOCS)*, pages 680–688. IEEE, 2007.
4. N. Bari and B. Pfitzmann. Collision-free accumulators and fail-stop signature schemes without trees. In *Advances in Cryptology — Eurocrypt '97*, volume 1233 of *LNCS*, pages 480–494. Springer, 1997.
5. M. Bellare and S. Micali. How to sign given any trapdoor function. In *Advances in Cryptology — Crypto '88*, volume 403 of *LNCS*, pages 200–215. Springer, 1990.
6. M. Bellare and S. Micali. How to sign given any trapdoor function. *Journal of the ACM*, 39(1):214–233, 1992.
7. M. Bellare and C. Namprempre. Authenticated encryption: Relations among notions and analysis of the generic composition paradigm. In *Advances in Cryptology — Asiacrypt 2000*, volume 1976 of *LNCS*, pages 531–545. Springer, 2000.
8. M. Bellare and T. Ristenpart. Simulation without the artificial abort: Simplified proof and improved concrete security for Waters' IBE scheme. In *Advances in Cryptology — Eurocrypt 2009*, volume 5479 of *LNCS*, pages 407–424. Springer, 2009.
9. M. Bellare and P. Rogaway. Random oracles are practical: A paradigm for designing efficient protocols. In *1st ACM Conference on Computer and Communications Security*, pages 62–73. ACM Press, 1993.
10. M. Bellare and P. Rogaway. The exact security of digital signatures: How to sign with RSA and Rabin. In *Advances in Cryptology — Eurocrypt '96*, volume 1070 of *LNCS*, pages 399–416. Springer, 1996.
11. M. Bellare and P. Rogaway. Collision-resistant hashing: Towards making UOWHFs practical. In *Advances in Cryptology — Crypto '97*, volume 1294 of *LNCS*, pages 470–484. Springer, 1997.
12. M. Bellare and S. Shoup. Two-tier signatures from the Fiat-Shamir transform, with applications to strongly unforgeable and one-time signatures. *IET Proc. Information Security*, 2(2):47–63, 2008.
13. D. J. Bernstein. Proving tight security for Rabin-Williams signatures. In *Advances in Cryptology — Eurocrypt 2008*, volume 4965 of *LNCS*, pages 70–87. Springer, 2008.

14. J. Black. The ideal-cipher model, revisited: An uninstantiable blockcipher-based hash function. In *Fast Software Encryption – FSE 2006*, volume 4047 of *LNCS*, pages 328–340. Springer, 2006.

15. D. Bleichenbacher and U. M. Maurer. On the efficiency of one-time digital signatures. In *Advances in Cryptology — Asiacrypt'96*, volume 1163 of *LNCS*, pages 145–158. Springer, 1996.

16. M. Blum. Coin flipping by telephone. In *Proc. IEEE Spring COMPCOM*, pages 133–137, 1982.

17. D. Boneh. Twenty years of attacks on the RSA cryptosystem. *Notices of the American Mathematical Society*, 46(2):203–213, 1999.

18. D. Boneh and X. Boyen. Short signatures without random oracles and the SDH assumption in bilinear groups. *Journal of Cryptology*, 21(2):149–177, 2008.

19. D. Boneh, B. Lynn, and H. Shacham. Short signatures from the Weil pairing. *Journal of Cryptology*, 17(4):297–319, September 2004.

20. J. N. Bos and D. Chaum. Provably unforgeable signatures. In *Advances in Cryptology — Crypto '92*, volume 740 of *LNCS*, pages 1–14. Springer, 1993.

21. E. F. Brickell, D. Pointcheval, S. Vaudenay, and M. Yung. Design validations for discrete logarithm based signature schemes. In *3rd Intl. Workshop on Theory and Practice in Public Key Cryptography(PKC 2000)*, volume 1751 of *LNCS*, pages 276–292. Springer, 2000.

22. D. R. L. Brown. Generic groups, collision resistance, and ECDSA. *Designs, Codes, and Cryptography*, 35(1):119–152, 2005.

23. C. Cachin, S. Micali, and M. Stadler. Computationally private information retrieval with polylogarithmic communication. In *Advances in Cryptology — Eurocrypt '99*, volume 1592 of *LNCS*, pages 402–414. Springer, 1999.

24. J. Camenisch and A. Lysyanskaya. A signature scheme with efficient protocols. In *3rd Intl. Conf. on Security in Communication Networks (SCN)*, volume 2576 of *LNCS*, pages 268–289. Springer, 2002.

25. R. Canetti, O. Goldreich, and S. Halevi. The random oracle methodology, revisited. *Journal of the ACM*, 51(4):557–594, 2004.

26. D. Catalano and R. Gennaro. Cramer-Damgård signatures revisited: Efficient flat-tree signatures based on factoring. In *8th Intl. Workshop on Theory and Practice in Public Key Cryptography(PKC 2005)*, volume 3386 of *LNCS*, pages 313–327. Springer, 2005.

27. B. Chevallier-Mames and M. Joye. A practical and tightly secure signature scheme without hash function. In *Cryptographers' Track — RSA 2007*, volume 4377 of *LNCS*, pages 339–356. Springer, 2007.

28. J.-S. Coron. On the exact security of full domain hash. In *Advances in Cryptology — Crypto 2000*, volume 1880 of *LNCS*, pages 229–235. Springer, 2000.

29. J.-S. Coron. Optimal security proofs for PSS and other signature schemes. In *Advances in Cryptology — Eurocrypt 2002*, volume 2332 of *LNCS*, pages 272–287. Springer, 2002.

30. J.-S. Coron and T. Icart. An indifferentiable hash function into elliptic curves. Available at http://eprint.iacr.org/2009/340.

31. J.-S. Coron and D. Naccache. Security analysis of the Gennaro-Halevi-Rabin signature scheme. In *Advances in Cryptology — Eurocrypt 2000*, volume 1807 of *LNCS*, pages 91–101. Springer, 2000.

32. R. Cramer. *Modular Design of Secure yet Practical Cryptographic Protocols*. PhD thesis, University of Amsterdam, 1996.

33. R. Cramer and I. Damgård. Secure signature schemes based on interactive protocols. In *Advances in Cryptology — Crypto '95*, volume 963 of *LNCS*, pages 297–310. Springer, 1995.

34. R. Cramer and I. Damgård. New generation of secure and practical RSA-based signatures. In *Advances in Cryptology — Crypto '96*, volume 1109 of *LNCS*, pages 173–185. Springer, 1996.

35. R. Cramer and T. Pedersen. Efficient and provable security amplifications. Technical Report CS-R9529, CWI, 1995.

36. R. Cramer and V. Shoup. Signature schemes based on the strong RSA assumption. *ACM Transactions on Information and System Security*, 3(3):161–185, 2000.

37. I. Damgård. Collision free hash functions and public key signature schemes. In *Advances in Cryptology — Eurocrypt '87*, volume 304 of *LNCS*, pages 203–216. Springer, 1988.

38. I. Damgård. A design principle for hash functions. In *Advances in Cryptology — Crypto '89*, volume 435 of *LNCS*, pages 416–427. Springer, 1990.

39. A. De Santis and M. Yung. On the design of provably secure cryptographic hash functions. In *Advances in Cryptology — Eurocrypt '90*, volume 473 of *LNCS*, pages 412–431. Springer, 1990.

40. W. Diffie and M. E. Hellman. New directions in cryptography. *IEEE Transactions on Information Theory*, 22(6):644–654, 1976.

41. Y. Dodis and L. Reyzin. On the power of claw-free permutations. In *3rd Intl. Conf. on Security in Communication Networks (SCN)*, volume 2576 of *LNCS*, pages 55–73. Springer, 2002.

42. C. Dwork and M. Naor. An efficient existentially unforgeable signature scheme and its applications. *Journal of Cryptology*, 11(3):187–208, 1998.

43. C. Dwork, M. Naor, O. Reingold, and L. Stockmeyer. Magic functions. *Journal of the ACM*, 50(6):852–921, 2003.

44. T. El Gamal. A public key cryptosystem and a signature scheme based on discrete logarithms. *IEEE Transactions on Information Theory*, 31:469–472, 1985.

45. S. Even, O. Goldreich, and S. Micali. On-line/off-line digital signatures. *Journal of Cryptology*, 9(1):35–67, 1996.

46. U. Feige, A. Fiat, and A. Shamir. Zero-knowledge proofs of identity. *Journal of Cryptology*, 1(2):77–94, 1988.

47. A. Fiat and A. Shamir. How to prove yourself: Practical solutions to identification and signature problems. In *Advances in Cryptology — Crypto '86*, volume 263 of *LNCS*, pages 186–194. Springer, 1987.

48. M. Fischlin. The Cramer-Shoup strong-RSA signature scheme revisited. In *6th Intl. Workshop on Theory and Practice in Public Key Cryptography(PKC 2003)*, volume 2567 of *LNCS*, pages 116–129. Springer, 2003.

49. M. Fischlin and R. Fischlin. The representation problem based on factoring. In *Cryptographers' Track — RSA 2002*, volume 2271 of *LNCS*, pages 96–113. Springer, 2002.

50. E. Fujisaki and T. Okamoto. Statistical zero knowledge protocols to prove modular polynomial relations. In *Advances in Cryptology — Crypto '97*, volume 1294 of *LNCS*, pages 16–30. Springer, 1997.

51. S. D. Galbraith, K. G. Paterson, and N. P. Smart. Pairings for cryptographers. *Discrete Applied Mathematics*, 156(16):3113–3121, 2008.

52. R. Gennaro, Y. Gertner, J. Katz, and L. Trevisan. Bounds on the efficiency of generic cryptographic constructions. *SIAM Journal on Computing*, 35(1):217–246, 2005.

53. R. Gennaro, S. Halevi, and T. Rabin. Secure hash-and-sign signatures without the random oracle. In *Advances in Cryptology — Eurocrypt '99*, volume 1592 of *LNCS*, pages 123–139. Springer, 1999.

54. E.-J. Goh, S. Jarecki, J. Katz, and N. Wang. Efficient signature schemes with tight reductions to the Diffie-Hellman problems. *Journal of Cryptology*, 20(4):493–514, 2007.

55. O. Goldreich. Two remarks concerning the Goldwasser-Micali-Rivest signature scheme. In *Advances in Cryptology — Crypto '86*, volume 263 of *LNCS*, pages 104–110. Springer, 1987.

56. O. Goldreich. *Foundations of Cryptography, vol. 1: Basic Tools*. Cambridge University Press, Cambridge, UK, 2001.

57. O. Goldreich. *Foundations of Cryptography, vol. 2: Basic Applications*. Cambridge University Press, Cambridge, UK, 2004.

58. S. Goldwasser and Y. Tauman Kalai. On the (in)security of the Fiat-Shamir paradigm. In *44th Annual Symposium on Foundations of Computer Science (FOCS)*, pages 102–115. IEEE, 2003.

59. S. Goldwasser and S. Micali. Probabilistic encryption. *Journal of Computer and System Sciences*, 28(2):270–299, 1984.

60. S. Goldwasser, S. Micali, and C. Rackoff. The knowledge complexity of interactive proof systems. *SIAM Journal on Computing*, 18(1):186–208, 1989.

61. S. Goldwasser, S. Micali, and R. L. Rivest. A digital signature scheme secure against adaptive chosen-message attacks. *SIAM Journal on Computing*, 17(2):281–308, 1988.
62. S. Goldwasser, S. Micali, and A. C.-C. Yao. Strong signature schemes. In *15th Annual ACM Symposium on Theory of Computing (STOC)*, pages 431–439. ACM Press, 1983.
63. L. C. Guillou and J.-J. Quisquater. A "paradoxical" indentity-based signature scheme resulting from zero-knowledge. In *Advances in Cryptology — Crypto '88*, volume 403 of *LNCS*, pages 216–231. Springer, 1990.
64. J. Håstad, R. Impagliazzo, L. A. Levin, and M. Luby. A pseudorandom generator from any one-way function. *SIAM Journal on Computing*, 28(4):1364–1396, 1999.
65. D. Hofheinz and E. Kiltz. Programmable hash functions and their applications. In *Advances in Cryptology — Crypto 2008*, volume 5157 of *LNCS*, pages 21–38. Springer, 2008.
66. S. Hohenberger and B. Waters. Realizing hash-and-sign signatures under standard assumptions. In *Advances in Cryptology — Eurocrypt 2009*, volume 5479 of *LNCS*, pages 333–350. Springer, 2009.
67. S. Hohenberger and B. Waters. Short and stateless signatures from the RSA assumption. In *Advances in Cryptology — Crypto 2009*, volume 5677 of *LNCS*, pages 654–670. Springer, 2009.
68. Q. Huang, D. S. Wong, and Y. Zhao. Generic transformation to strongly unforgeable signatures. In *ACNS 07: 5th International Conference on Applied Cryptography and Network Security (ACNS)*, volume 4521 of *LNCS*, pages 1–17. Springer, 2007.
69. M. Jakobsson. Reducing costs in identification protocols. Presented at the rump session of Crypto 1992. Available at http://www.informatics.indiana.edu/markus/papers.asp.
70. M. Joye. How (not) to design strong-RSA signatures. *Designs, Codes, and Cryptography*. To appear.
71. J. Katz and C.-Y. Koo. On constructing universal one-way hash functions from arbitrary one-way functions. Available at http://eprint.iacr.org/2005/328.
72. J. Katz and Y. Lindell. *Introduction to Modern Cryptography*. Chapman & Hall/CRC Press, 2007.
73. J. Katz and N. Wang. Efficiency improvements for signature schemes with tight security reductions. In *ACM CCS '03: 10th ACM Conference on Computer and Communications Security*, pages 155–164. ACM Press, 2003.
74. H. Krawczyk and T. Rabin. Chameleon signatures. In *Network and Distributed System Security Symposium – NDSS 2000*. The Internet Society, 2000.
75. K. Kurosawa and K. Schmidt-Samoa. New online/offline signature schemes without random oracles. In *9th Intl. Conference on Theory and Practice of Public Key Cryptography(PKC 2006)*, volume 3958 of *LNCS*, pages 330–346. Springer, 2006.
76. L. Lamport. Constructing digital signatures from a one-way function. Technical Report SRI-CSL-98, SRI Intl. Computer Science Laboratory, October 1979.
77. J. Malone-Lee and N. P. Smart. Modifications of ECDSA. In *SAC 2002: 9th Annual International Workshop on Selected Areas in Cryptography (SAC)*, volume 2595 of *LNCS*, pages 1–12. Springer, 2003.
78. R. C. Merkle. Protocols for public key cryptosystems. In *IEEE Symposium on Security & Privacy*, pages 122–134. IEEE, 1980.
79. R. C. Merkle. A digital signature based on a conventional encryption function. In *Advances in Cryptology — Crypto '87*, volume 293 of *LNCS*, pages 369–378. Springer, 1988.
80. R. C. Merkle. A certified digital signature (that antique paper from 1979). In *Advances in Cryptology — Crypto '89*, volume 435 of *LNCS*, pages 218–238. Springer, 1990.
81. R. C. Merkle. One way hash functions and DES. In *Advances in Cryptology — Crypto '89*, volume 435 of *LNCS*, pages 428–446. Springer, 1990.
82. S. Micali. A secure and efficient digital signature algorithm. Technical Report MIT/LCS/TM-501b, Massachusetts Institute of Technology, Laboratory for Computer Science, April 1994.
83. S. Micali, M. O. Rabin, and S. P. Vadhan. Verifiable random functions. In *40th Annual Symposium on Foundations of Computer Science (FOCS)*, pages 120–130. IEEE, 1999.
84. S. Micali and L. Reyzin. Improving the exact security of digital signature schemes. *Journal of Cryptology*, 15(1):1–18, 2002.

85. M. Mitzenmacher and A. Perrig. Bounds and improvements for BiBa signature schemes. Technical Report TR-02-02, Harvard University, 2002.

86. D. Naccache, D. Pointcheval, and J. Stern. Twin signatures: An alternative to the hash-and-sign paradigm. In *ACM CCS '01: 8th ACM Conference on Computer and Communications Security*, pages 20–27. ACM Press, 2001.

87. M. Naor. On cryptographic assumptions and challenges (invited talk). In *Advances in Cryptology — Crypto 2003*, volume 2729 of *LNCS*, pages 96–109. Springer, 2003.

88. M. Naor and M. Yung. Universal one-way hash functions and their cryptographic applications. In *21st Annual ACM Symposium on Theory of Computing (STOC)*, pages 33–43. ACM Press, 1989.

89. J. B. Nielsen. Separating random oracle proofs from complexity theoretic proofs: The non-committing encryption case. In *Advances in Cryptology — Crypto 2002*, volume 2442 of *LNCS*, pages 111–126. Springer, 2002.

90. National Institute of Standards and Technology. Digital signature standard (DSS). Federal Information Processing Standards (FIPS) Publication #186-3, 2009. Available at `http://www.itl.nist.gov/fipspubs/by-num.htm`.

91. K. Ohta and T. Okamoto. A modification of the Fiat-Shamir scheme. In *Advances in Cryptology — Crypto '88*, volume 403 of *LNCS*, pages 232–243. Springer, 1990.

92. K. Ohta and T. Okamoto. On concrete security treatment of signatures derived from identification. In *Advances in Cryptology — Crypto '98*, volume 1462 of *LNCS*, pages 354–369. Springer, 1998.

93. T. Okamoto. Provably secure and practical identification schemes and corresponding signature schemes. In *Advances in Cryptology — Crypto '92*, volume 740 of *LNCS*, pages 31–53. Springer, 1993.

94. H. Ong and C.-P. Schnorr. Fast signature generation with a Fiat-Shamir-like scheme. In *Advances in Cryptology — Eurocrypt '90*, volume 473 of *LNCS*, pages 432–440. Springer, 1990.

95. PKCS #1 version 2.1: RSA cryptography standard. RSA Data Security, Inc., 1998. Available at `http://www.rsa.com/rsalabs`.

96. D. Pointcheval and J. Stern. Security arguments for digital signatures and blind signatures. *Journal of Cryptology*, 13(3):361–396, 2000.

97. M. O. Rabin. Digitalized signatures and public-key functions as intractable as factorization. Technical Report MIT/LCS/TR-212, Laboratory for Computer Science, MIT, January 1979.

98. L. Reyzin and N. Reyzin. Better than BiBa: Short one-time signatures with fast signing and verifying. In *7th Australian Conference on Information Security and Privacy, ACISP 2002*, volume 2384 of *LNCS*, pages 144–153. Springer, 2002.

99. R. L. Rivest, A. Shamir, and L. M. Adleman. A method for obtaining digital signatures and public-key cryptosystems. *Communications of the ACM*, 21(2):120–126, 1978.

100. J. Rompel. One-way functions are necessary and sufficient for secure signatures. In *22nd Annual ACM Symposium on Theory of Computing (STOC)*, pages 387–394. ACM Press, 1990.

101. C.-P. Schnorr. Efficient signature generation by smart cards. *Journal of Cryptology*, 4(3):161–174, 1991.

102. A. Shamir. On the generation of cryptographically strong pseudorandom sequences. *ACM Trans. on Computer Systems*, 1(1):38–44, 1983.

103. A. Shamir and Y. Tauman. Improved online/offline signature schemes. In *Advances in Cryptology — Crypto 2001*, volume 2139 of *LNCS*, pages 355–367. Springer, 2001.

104. V. Shoup. On the security of a practical identification scheme. *Journal of Cryptology*, 12(4):247–260, 1999.

105. V. Shoup. A composition theorem for universal one-way hash functions. In *Advances in Cryptology — Eurocrypt 2000*, volume 1807 of *LNCS*, pages 445–452. Springer, 2000.

106. R. Steinfeld, J. Pieprzyk, and H. Wang. How to strengthen any weakly unforgeable signature into a strongly unforgeable signature. In *Cryptographers' Track — RSA 2007*, volume 4377 of *LNCS*, pages 357–371. Springer, 2007.

107. S. Vaudenay. The security of DSA and ECDSA. In *6th Intl. Workshop on Theory and Practice in Public Key Cryptography(PKC 2003)*, volume 2567 of *LNCS*, pages 309–323. Springer, 2003.
108. L. Washington. *Elliptic Curves: Number Theory and Cryptography*. CRC Press, 2008.
109. B. R. Waters. Efficient identity-based encryption without random oracles. In *Advances in Cryptology — Eurocrypt 2005*, volume 3494 of *LNCS*, pages 114–127. Springer, 2005.
110. H. C. Williams. A modification of the RSA public-key encryption procedure. *IEEE Transactions on Information Theory*, 26(6):726–729, 1980.

Index